NORTH CAROLINA
STATE BOARD OF COMMUNITY COLLEGES
LIBRARIES
ASHEVILLE-BUNCOMBE TECHNICAL CC

DISCARDED

NOV 20 2024

A REVIEW OF THE STRUCTURE AND PHYSICAL PROPERTIES OF LIQUID CRYSTALS

Authors:

GLENN H. BROWN
Liquid Crystal Institute
Kent State University

J. W. DOANE
Liquid Crystal Institute and
Department of Physics
Kent State University

VERNON D. NEFF
Liquid Crystal Institute and
Department of Chemistry
Kent State University

published by:

A DIVISION OF
THE **CHEMICAL RUBBER** CO.
18901 Cranwood Parkway • Cleveland, Ohio 44128

This book represents information obtained from authentic and highly regarded sources. Reprinted material is quoted with permission, and sources are indicated. A wide variety of references is listed. Every reasonable effort has been made to give reliable data and information, but the author and the publisher cannot assume responsibility for the validity of all materials or for the consequences of their use.

Copyright © The Chemical Rubber Co. 1971

CRC MONOTOPIC SERIES

The primary objective of the CRC Monotopic Series is to provide reference works, each of which represents an authoritative and comprehensive summary of the "state-of-the-art" of a single well-defined scientific subject.

Among the criteria utilized for the selection of the subject are: (1) timeliness; (2) significant recent work within the area of the subject; and (3) recognized need of the scientific community for a critical synthesis and summary of the "state-of-the-art."

The value and authenticity of the contents are assured by utilizing the following carefully structured procedure to produce the final manuscript:

1. The topic is selected and defined by an editor and advisory board, each of whom is a recognized expert in the discipline.

2. The author, appointed by the editor, is an outstanding authority on the particular topic which is the subject of the publication.

3. The author, utilizing his expertise within the specialized field, selects for critical review the most significant papers of recent publication and provides a synthesis and summary of the "state-of-the-art."

4. The author's manuscript is critically reviewed by a referee who is acknowledged to be equal in expertise in the specialty which is the subject of the work.

5. The editor is charged with the responsibility for final review and approval of the manuscript.

In establishing this new CRC Monotopic Series, CRC has the additional objective of attacking the high cost of publishing in general, and scientific publishing in particular. By confining the contents of each book to an *in-depth treatment* of a relatively narrow and well-defined subject, the physical size of the book, itself, permits a pricing policy substantially below current levels for scientific publishing.

Although well-known as a publisher, CRC now prefers to identify its function in this area as the management and distribution of scientific information, utilizing a variety of formats and media ranging from the conventional printed page to computerized data bases. Within the scope of this framework, the CRC Monotopic Series represents a significant element in the total CRC scientific information service.

B. J. Starkoff, President
THE CHEMICAL RUBBER Co.

This book originally appeared as part of an article in *CRC Critical Reviews in Solid State Sciences*, a quarterly journal published by the Chemical Rubber Co. We would like to acknowledge the editorial assistance received by the journal's editors, Donald E. Schuele, Ph.D., Bell Telephone Laboratories, Columbus, and Richard W. Hoffman, Ph.D., Case Western Reserve University, Cleveland. G. W. Gray, Ph.D., University of Hull, East Yorkshire, England, served as referee for this article.

AUTHORS' INTRODUCTION

It is obviously impossible to give a complete survey of the field of liquid crystals in the few pages at our disposal. The material presented herein is a review and not a textbook or treatise on the subject of liquid crystals. Some topics had to be omitted and some significant areas are discussed only casually. We have, therefore, chosen to review selected, recent literature on the structure and physical properties of thermotropic liquid crystals and to include a brief review of lyotropic systems.

In writing this review article on the "state of the art" in liquid crystals, we assume that the reader has an acquaintance with the fundamentals of the subject. For those who may wish to start with more elementary and general material, adequate references are cited. In addition to the references cited to specific articles, supplemental materials in the way of review articles and books are brought to the attention of the reader.

The main focus of the review is on the literature from January 1960 to February 1970 which is found in abstract and scientific journals. We have included selected articles since the February date and have also included in Appendix A a list of the papers which were presented at the Third International Liquid Crystal Conference in Berlin, Germany, August 24-28, 1970. In Appendix B we have introduced new material on the subject of diffusion in liquid crystals. Most of the papers presented at the conference are appearing in current issues of *Molecular Crystals and Liquid Crystals* and will be collected later in a hardback volume.

Glenn H. Brown
J. W. Doane
Vernon D. Neff
Kent, Ohio

THE AUTHORS

Glenn H. Brown is Director of the Liquid Crystal Institute and Regents Professor, Kent State University.

Dr. Brown received his B.S. degree from Ohio University in 1939, his M.S. degree from Ohio State University in 1941, and his Ph.D. degree from Iowa State University in 1951.

J. W. Doane is Associate Professor of Physics and associated with the Liquid Crystal Institute, Kent State University.

Dr. Doane received his B.S., M.S., and Ph.D. degrees from the University of Missouri. He has publications in the field of liquid crystals.

Vernon D. Neff is Associate Professor of Chemistry and associated with the Liquid Crystal Institute, Kent State University.

Dr. Neff received his B.S. degree in 1953 and Ph.D. degree in Chemical Physics in 1960 from Syracuse University.

He has authored a number of papers in his major areas of research including liquid crystals, infrared and Raman spectroscopy, and theoretical chemistry.

TABLE OF CONTENTS

I.	Molecular Geometry of Molecules Which Form Thermotropic Liquid Crystals	9
II.	Polymorphism in Thermotropic Liquid Crystals	12
III.	Structure of Thermotropic Liquid Crystals	15
IV.	Theories of Liquid Crystalline Structures	21
	A. The Swarm Theory of Nematic Liquid Crystals	21
	B. The Continuum Theory	22
	C. Curvature Strains in Liquid Crystals	23
	D. Nematic and Cholesteric Structures	23
	E. The Smectic Phase	26
V.	The Effect of External Forces and Fields	28
	A. Nematic Materials	28
	B. Cholesteric Materials	33
	C. Smectic Materials	35
VI.	Thermodynamics of Thermotropic Liquid Crystals	35
	A. Statistical Theory of Phase Transitions	36
	B. Thermodynamic Measurements	39
VII.	Infrared and Raman Spectra of Liquid Crystals	42
VIII.	Light Scattering and Spin-Lattice Relaxation	42
IX.	Magnetic Resonance	49
	A. Nuclear Magnetic Resonance	50
	B. Electron Paramagnetic Resonance	56
	C. Mossbauer Effects	62
X.	Other Physical Properties	63
	A. Viscosity	63
	B. Ultrasonics	64
	C. Brillouin Scattering	64
	D. Positron Annihilation	64
	E. Birefringence	64
XI.	Lyotropic Liquid Crystals	65
	A. Amphiphilic Compounds (Amphiphiles)	65
	B. Micelles	65
	C. Classes of Lyotropic Structures	66
	D. Composition of a Typical System and Occurence of Various Liquid Crystalline Phases in that System	73
	E. Transitions Between Different Lyotropic Liquid Crystal Phases	73
	F. Summary	76
	Acknowledgments	76
	Appendix A	85
	Appendix B	94
	References	85

In writing this review article on the "state of the art" in liquid crystals we assume that the reader has an acquaintance with the fundamentals of the subject. For those who may wish to start with more elementary and general material we cite two articles.[1,2]

An effort has been made to give a critical review of the literature since 1960, covering the most important developments on the subject of the paper. In addition to the references cited to specific articles, supplemental materials in the way of review articles[3-7] and books[8-11] are available to the interested reader. A listing of thermotropic liquid crystals can be found in Landolt-Börnstein.[13] Even though the listing was published in 1960, it is still the best available.

Lehmann[12] was the first to suggest the name *liquid crystals* to identify the state of matter discussed in this paper. The term *liquid crystal* has been used by researchers in the field since Lehmann's suggestion although many other terms have been proposed. Such terms as *mesomorphs* or *mesoforms*, *mesomorphic states* and *paracrystals* have been proposed and used rather freely in the literature. More recently one finds such terms as *anisotropic liquids, anisotropic fluids,* and for specific classes of liquid crystals such expressions as *nematic liquid, smectic liquid* and *cholesteric liquid*.

It is obviously impossible to give a complete survey of the field of liquid crystals in the few pages at our disposal. We have, therefore, chosen to cover the major aspects of the structure and physical properties of thermotropic liquid crystals. A brief review of lyotropic systems is included.

I. MOLECULAR GEOMETRY OF MOLECULES WHICH FORM THERMOTROPIC LIQUID CRYSTALS

The kinds of molecules which form liquid crystals generally possess certain features of common geometry even though the compounds may be of a variety of chemical types such as anils, azo compounds, azoxy compounds or cholesteryl esters. The molecular features which one finds in compounds which form thermotropic liquid crystals may be summarized as follows:

1. The molecules will be elongated and rectilinear. If the molecule has "flat" segments, e.g. benzene rings, liquid crystallinity will be enhanced.

2. The molecule will be "rigid" along its long axis; double bond(s) are common along this axis of the molecule.

3. Simultaneous existence of strong dipoles and easily polarizable groups in the molecule seems important. The most pronounced liquid crystallinity effect is most likely to occur if the strong dipole is on the molecular axis.

4. Weak dipolar groups at the extremities of the molecule are of subordinate importance.

The kind of fragments one uses "to build" thermotropic liquid crystalline compounds are illustrated in Table 1. Specific molecules, using

TABLE 1

Examples of Atomic and Molecular Fragments from which Thermotropic Liquid Crystals may be Built

(a) Aromatic Groups

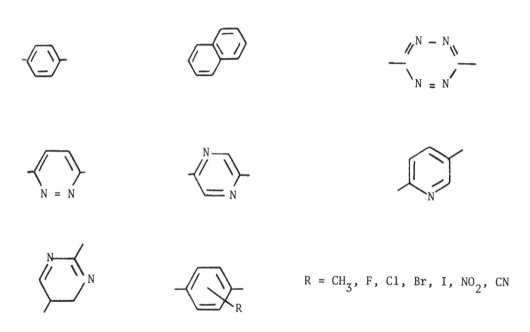

(b) Linkage Groups

$-(CH_2)_n-$	$-O-(CH_2)_n-$	$-CH_2-NH-$	$-CH = N-$ $\quad\quad\quad\downarrow$ $\quad\quad\quad O$
$-N = N-$	$-N = N-$ $\quad\quad\downarrow$ $\quad\quad O$	$-CH = CH-$	$-Hg-$
$-CH = N-$		$-S-(CH_2)_n-$	$-C \equiv C-$

(c) Other Cyclic Groups

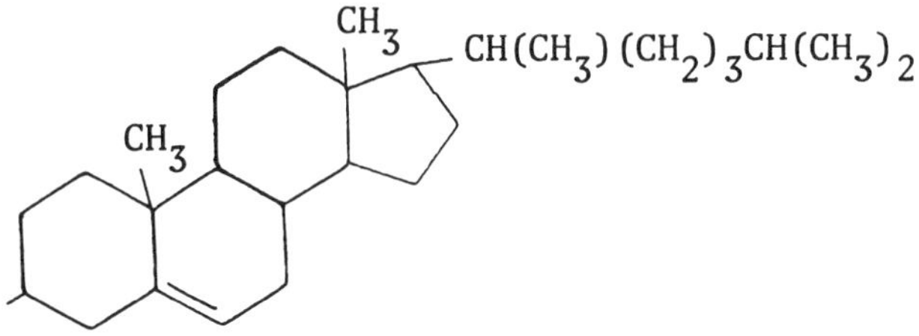

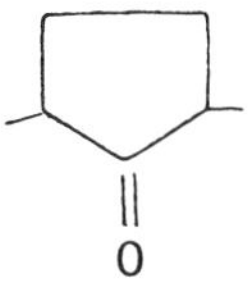

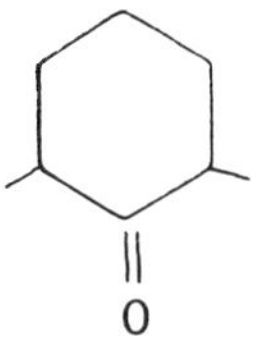

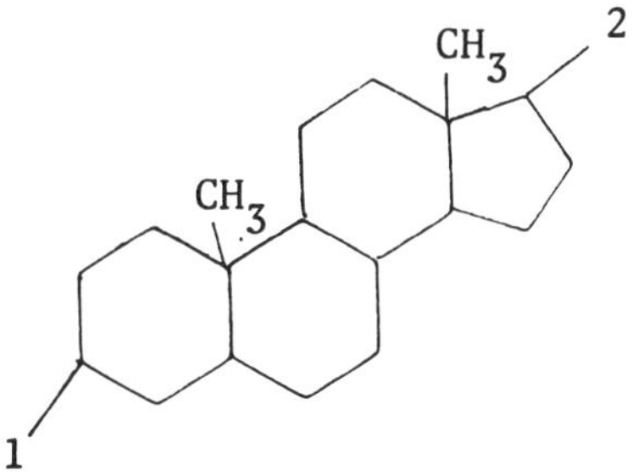

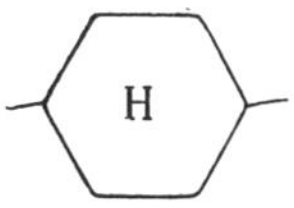

transform

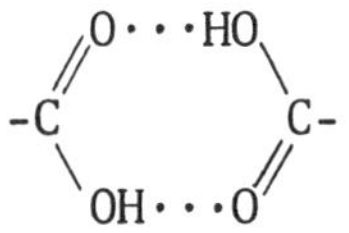

Groups in positions 1 and 2 may vary.
The degree and position in the ring
are changeable.

(d) Terminal Groups

$CH_3(CH_2)_n$-

RO-   R = normal alkyl or aromatic group

F, Cl, Br, I-

NC-

O_2N-

(d) Terminal Groups (Continued)

H$_2$N-

Me$_2$N-

RCOO-   R = normal alkyl or aromatic group

$$R-O-\overset{O}{\underset{\|}{C}}-$$
R = alkyl or aromatic group, or Li, Na, K, Rb, Cs, Tl

CH$_3$-O-(CH$_2$)$_n$-O-

these fragments as building units, are shown in Table 2 along with transition temperatures at phase changes. Neither of these tables is meant to be exhaustive in its contents.

The information recorded in Tables 1 and 2 indicates that aromatic types of molecules are most prominent in thermotropic liquid crystals. However, we should not ignore aliphatic compounds even though the number of such compounds forming liquid crystals is small compared to the aromatic type. The monocarboxylic acids and their salts are the most common aliphatic compounds showing liquid crystallinity. The simplest aliphatic compound that shows liquid crystallinity is 2,4-nonadienoic acid.

Recent advances in the synthesis of liquid crystals and the role of molecular geometry in liquid crystallinity have been described effectively by Gray.[14]

II. POLYMORPHISM IN THERMOTROPIC LIQUID CRYSTALS

Polymorphism in thermotropic liquid crystalline systems has been well documented during the last five years. The main classes of liquid crystals are identified as nematic, cholesteric (a twisted nematic) and smectic textures. Sackmann et al.[7] have identified five smectic textures. The compilation in Table 3 is based on the work of these authors.

A number of observations can be made on analysis of Table 3. Some of these are as follows:

1. If a substance exhibits both nematic and smectic states, the nematic state is the higher temperature form.

2. If a substance shows smectic trimorphous textures, the temperature relationships are in this order:

Isotropic → Nematic → Smectic A → Liquid
$$\xrightarrow{\text{Smetic C → Smectic B}}$$
Decreasing Temperature

3. Compounds with smectic C textures may exhibit this texture alone or together with a nematic or a smectic A texture. The smectic C modification is the lower temperature one.

4. In the presence of smectic A and/or smectic C, the smectic B modification is always the lower temperature one.

It has been generally accepted that there is only one nematic texture exclusive of the cholesteric texture (twisted nematic). de Vries[28] has recently presented evidence for the existence of more than one nematic texture. de Vries has proposed for one of these textures a skewed cybotactic phase, in which the molecules are arranged in groups in such a way that the centers of the molecules in each group

TABLE 2

Typical Thermotropic Liquid Crystalline Compounds and their Transition Temperatures

1. 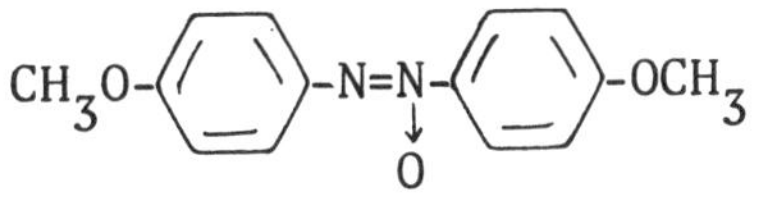 CH₃O-⟨⟩-N=N(→O)-⟨⟩-OCH₃

   p-azoxyanisole

   Solid ←118°→ Nematic Liquid ←136°→ Isotropic Liquid[13]

2. 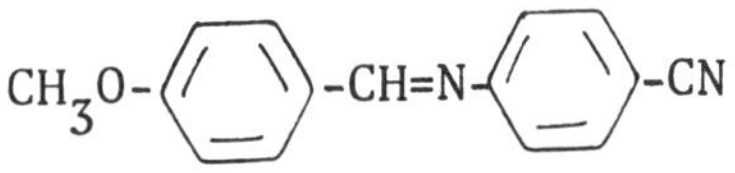 CH₃O-⟨⟩-CH=N-⟨⟩-CN

   4-methoxybenzylidene-4'-cyanoaniline

   Solid ←107°→ Nematic Liquid ←120°→ Isotropic Liquid[15]

3. 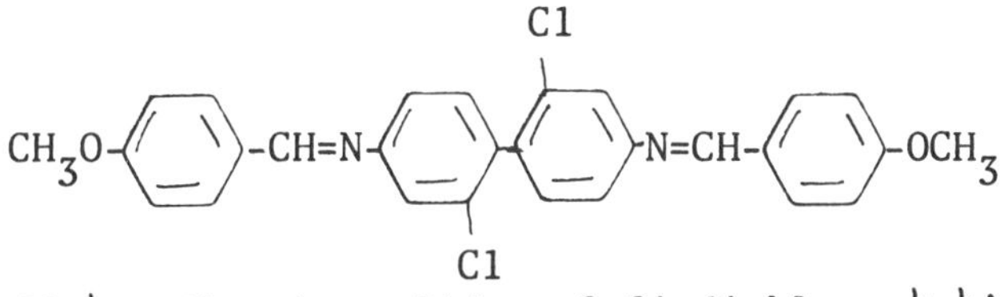 CH₃O-⟨⟩-CH=N-⟨⟩(Cl)-⟨⟩(Cl)-N=CH-⟨⟩-OCH₃

   di-4-methoxybenzylidene-2,2'-dichloro-4,4'-diaminobiphenyl

   Solid ←216°→ Nematic Liquid ←321°→ Isotropic Liquid[16]

4. 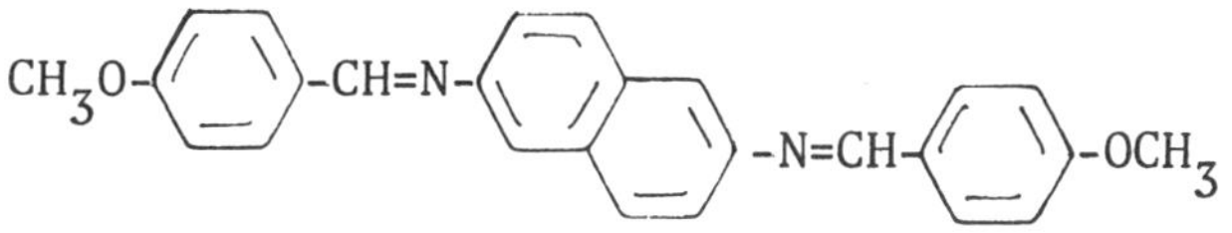 CH₃O-⟨⟩-CH=N-⟨naphthalene⟩-N=CH-⟨⟩-OCH₃

   di-4-methoxybenzylidene-2,6-diaminonaphthalene

   Solid ←189°→ Nematic Liquid ←356°→ Isotropic Liquid[16]

5. 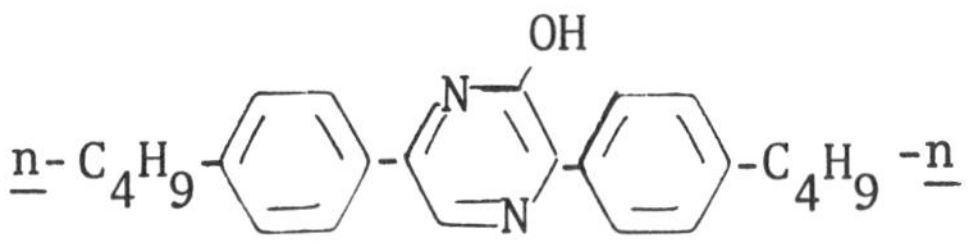 n-C₄H₉-⟨⟩-⟨pyrazine-OH⟩-⟨⟩-C₄H₉-n

   2-hydroxy-3,6-bis(4-n-butylphenyl)-pyrazine

   Solid ←247°→ Smectic Liquid ←300°→ Isotropic Liquid[17]

6.

p-quinquephenyl

Solid $\xleftrightarrow{401°}$ Nematic Liquid $\xleftrightarrow{445°}$ Isotropic Liquid[18]

7. 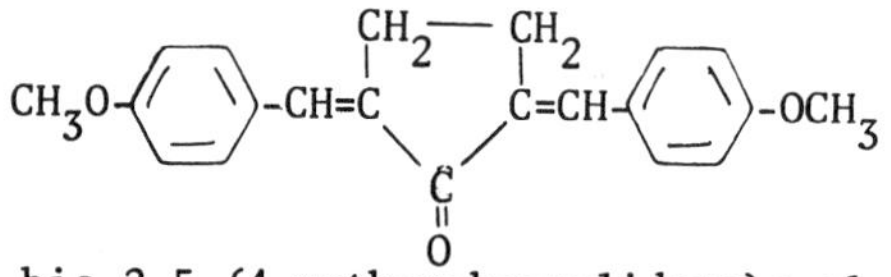 $CH_3O-\bigcirc-CH=C\underset{\underset{O}{C}}{\overset{CH_2-CH_2}{\diagup\diagdown}}C=CH-\bigcirc-OCH_3$

bis-2,5-(4-methoxybenzylidene)cyclopentanone

Nematic Liquid $\xleftrightarrow{195°}$ Isotropic Liquid[19]
↓ ↑
Solid $\xleftarrow{216°}$

8.

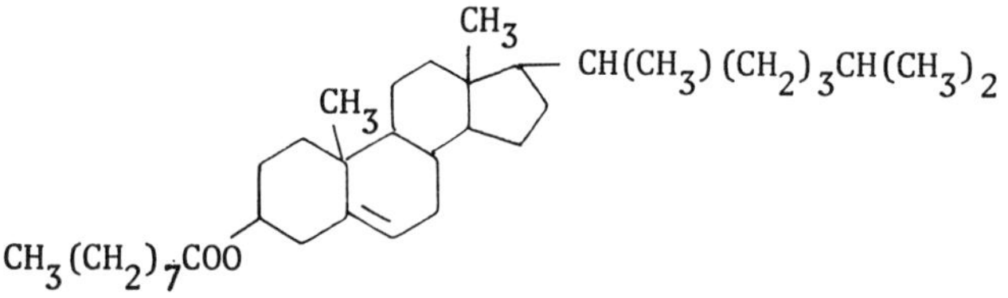

cholesteryl-n-nonanoate

Solid $\xleftrightarrow{74°}$ Smectic Liquid $\xleftrightarrow{76.3°}$ Cholesteric Liquid $\xleftrightarrow{92.1°}$

Isotropic Liquid[20]

9.

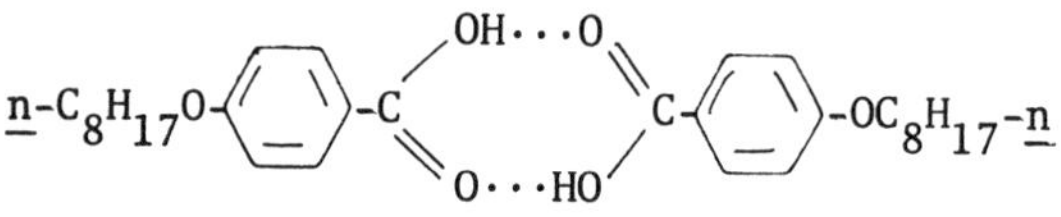

4-n-octyloxybenzoic acid (dimer)

Solid 2 $\xleftrightarrow{71.1°}$ Solid 1 $\xleftrightarrow{100.7°}$ Smectic Liquid $\xleftrightarrow{107.5°}$

Isotropic Liquid $\xleftrightarrow{147.3°}$ Nematic Liquid[21]

10. CH₃O-⟨phenyl⟩-O-C(=O)-⟨H-cyclohexane-H⟩-C(=O)-O-⟨phenyl⟩-OCH₃

di-4-methoxyphenyl <u>trans</u> -1,4-cyclohexanedicarboxylate

Solid ⟵143°⟶ Nematic Liquid ⟵242°⟶ Isotropic Liquid [22]

11. <u>n</u>-C₄H₉O-⟨phenyl⟩-N=N-⟨phenyl⟩-COCH₃

4-<u>n</u>-butyloxy-4'-acetylazobenzene

Solid ⟵109.1°⟶ Smectic Liquid ⟵112.5°⟶ Nematic Liquid ⟵122.2°⟶ Isotropic Liquid [23]

12. <u>n</u>-C₆H₁₃-O-⟨phenyl⟩-CH=N-⟨phenyl⟩-Hg-⟨phenyl⟩-N=CH-⟨phenyl⟩-O-C₆H₁₃-<u>n</u>

<u>bis</u> [N-(4-<u>n</u>-hexyloxybenzylidene)-4'-aminophenyl] mercury

Solid ⟵151°⟶ Smectic Liquid ⟵193°⟶ Smectic Liquid ⟵197°⟶ Smectic Liquid ↕290°

Isotropic Liquid ⟵333°⟶ Nematic Liquid [24]

lie in a plane making an angle ϕ significantly different from 90° with the mean direction of the long axes of the molecules in that group. A second texture has an arrangement which de Vries identifies as the normal cybotactic phase and it is similar to the skewed one but with ϕ close to 90°. The third texture proposed by de Vries is the classical nematic phase in which no regular arrangement between neighboring molecules exists. In this phase the only molecular pattern is the essentially parallel ordering of the long axes of the molecules to one another.

III. STRUCTURE OF THERMOTROPIC LIQUID CRYSTALS

Studies of the structures of thermotropic liquid crystals have been attacked by several different treatments of X-ray data. One approach involves taking Laue and/or Debye-Scherrer patterns of liquid crystalline textures. These patterns are then analyzed to establish information on intermolecular arrangements, strata packing or other structural facts. A second approach is treatment of the diffraction data by the use of radial distribution functions (atomic or molecular) or by use of a cylindrical coordinate system provided the molecular orientations are proper. A third approach is determination of crystal structures of compounds which form thermotropic liquid crystals. All three kinds of studies are in progress in the Liquid Crystal Institute.

We shall first consider the information on structure gained from X-ray patterns of the Laue type of the different nematic and smectic systems. If there is no overall preferred orientation of the molecules, all three types of nematic systems described by de Vries[28] generally give diffraction patterns which have two diffraction rings. The inner ring is related to the length of the molecule and the outer ring to the average distance between neighboring molecules. By use of a magnetic field or some other external force, the molecules can be

TABLE 3

Classification of Smectic Liquid Crystalline Compounds Based on Miscibility Studies of Binary Systems

Type	Transition Temperatures Between Indicated Textures					Example
	T_0	T_1	T_2	T_3		
	Texture	Texture	Texture	Texture	Texture	
a	i	n				$CH_3O-\langle\rangle-N=N-\langle\rangle-OCH_3$ (↑O) p-azoxyanisole[13]
	i	chol				cholesteryl acetate[13]
b_1	i	S_A				$CH_2=CHCH_2OOC-\langle\rangle-N=N-\langle\rangle-COOCH_2CH=CH_2$ (↓O) diallyl azoxybenzene 4,4' dicarboxylate[23]
b_2	i	S_C				$C_{16}H_{33}OOC-C=CH-\langle\rangle-N=N-\langle\rangle-CH=C-COOC_{16}H_{33}$ (↑O), CH_3 groups di-n-hexadecyl 4,4'-azoxy-α-methylcinnamate[23]
c_1	i	n	S_A			$Br-\langle\rangle-COO-\langle\rangle-N=N-\langle\rangle-COOC_2H_5$ (4-carbethoxyphenylazophenyl) 4'-bromobenzoate[23]
	i	chol	S_A			cholesteryl nonanoate[20]
c_2	i	n	S_B			$C_3H_7S-\langle\rangle-CH=N-\langle\rangle-N=N-\langle\rangle$ 4-n-propylmercaptobenzal-4-aminoazobenzene
c_3	i	n	S_C			$n-C_8H_{17}O-\langle\rangle-COOH$ n-octyloxybenzoic acid[23]
d_1	i	S_A	S_B			$C_2H_5O-\langle\rangle-CH=N-\langle\rangle-CH=CHCOOC_2H_5$ ethyl 4-ethoxybenzal-4'-aminocinnamate[23]

Key:
T_0 Liquid crystal-isotropic liquid transition temperature.
T_1, T_2, T_3 Transition temperatures between liquid crystalline modifications.
i   Isotropic liquid.
n   Nematic texture.
chol   Cholesteric texture.
S_A, S_B, S_C, S_D, S_E Smectic textures with special characteristics.

Type	Transition Temperatures Between Indicated Textures					Example
	T_0 Texture	T_1 Texture	T_2 Texture	T_3 Texture	Texture	
d_2	i	S_A	S_C			iso-amyl-OOCCH=CH-⟨○⟩-N=N-⟨○⟩-CH=CHCOO-iso amyl, ↓O diisoamyl 4,4'-**azoxycinnamate**[23]
d_3	i	S_A	S_E			$\underline{n}$-C$_3$H$_7$OOC-⟨○⟩-⟨○⟩-⟨○⟩-COOC$_3$H$_7$-$\underline{n}$ di-$\underline{n}$-propyl 4-terphenyl-4,4''-dicarboxylate[25]
d_4	i	S_C	S_B			$\underline{n}$-C$_{18}$H$_{37}$O-⟨○⟩-N=N-⟨○⟩-OC$_{18}$H$_{37}$-$\underline{n}$, ↑O 4,4'-$\underline{n}$-**octadecyloxyazobenzene**
d_5	i	S_D	S_C			$\underline{n}$-C$_{18}$H$_{37}$O-⟨○⟩-⟨○⟩-COOH, NO$_2$ 4'-$\underline{n}$-octadecyloxy-3'-**nitrobiphenyl**-4-carboxylic acid[25]
e_1	i	n	S_A	S_B		C$_2$H$_5$O-⟨○⟩-CH=N-⟨○⟩-CH=CHCOOC$_2$H$_5$ ethyl 4-ethoxybenzal-4'-**aminocinnamate**[23]
e_2	i	n	S_A	S_C		$\underline{n}$-C$_6$H$_{13}$-O-⟨○⟩-⟨○⟩-COOH, NO$_2$ 4'-$\underline{n}$-hexyloxy-3'-**nitrobiphenyl**-4-carboxylic acid[26]
e_3	i	n	S_C	S_B		C$_6$H$_{13}$O-⟨○⟩-⟨N◯N⟩-⟨○⟩-OC$_6$H$_{13}$ 2,5-$\underline{bis}$-(4-$\underline{n}$-hexyloxyphenyl)pyrazine
e_4	i	S_A	S_C	S_B		$\underline{n}$C$_{10}$H$_{21}$O-⟨○⟩-CH=N-⟨○⟩-CH=CH-C-O $\underline{n}$-C$_5$H$_{11}$, ‖O $\underline{n}$-amyl-4-$\underline{n}$-decyloxybenzal-4'-**aminocinnamate**[27]
e_5	i	S_A	S_D	S_C		$\underline{n}$-C$_{16}$H$_{33}$O-⟨○⟩-⟨○⟩-COOH, NO$_2$ 4'-$\underline{n}$-hexadecyloxy-3'-**nitrobiphenyl**-4-carboxylic acid[26]
g	i	n	S_A	S_C	S_B	C$_2$H$_5$O-C-CH=CH-⟨○⟩-N=CH-⟨○⟩-**CH=N**-⟨○⟩-CH=CH-C-OC$_2$H$_5$, ‖O ‖O diethyl **terephthal-bis-**(4-aminocinnamate)[27]

aligned in a direction perpendicular to the incident beam and the skewed cybotactic phase which exists in the system has a diffraction pattern which shows four maxima in the inner ring in the directions making an angle ϕ with the direction in which the two maxima are oriented in the outer ring. The angle ϕ is the complement of the skew angle. The classical molecular arrangement and the normal cybotactic phase show two maxima[28] in the inner ring directed perpendicular to the two maxima in the outer ring (also, see Falgueirettes[30]). de Vries[28] has pointed out that X-ray diffraction patterns of the normal cybotactic and classical phases will not be greatly different but, nevertheless, sufficiently different that they should be noticeable. Comparing the two diffraction patterns one should find that the normal cybotactic phase gives a sharper inner ring than the classical phase and the diameter of this ring is not dependent on the degree of preferred orientation in the normal cybotactic phase but is dependent for the independent molecules (classical phase).

Sackmann and his co-workers[7,25] have done some interesting work on classifying smectic textures. They found that S_A, S_B and S_C phases have certain intensity versus angle diffraction profiles which are similar. de Vries[28] and Chistyakov et al.[32b] have shown by X-ray studies that S_C has a tilted arrangement of molecules within a stratum. The plot of intensity versus angle for the S_A and S_B textures is presented in Figure 1. The sharp interferences at small angles in both the smectic A and smectic B textures correspond to the spacing between strata. In smectic A, the broad peak at about 10° may be attributed primarily to the statistical distribution of the molecules in the layers, the long axes of the molecules being normal to the planes of the strata. This reflection indicates that the molecules in smectic A have essentially a random distribution within a stratum. For smectic B there is a sharp peak at 10°, indicating that there is much more order in the arrangement of molecules in a stratum than is found in smectic A. The X-ray diffraction pattern of the smectic C modification has the sharp peak at the small angle similar to the findings for smectic A and smectic B. The profile of the peak at 10°

FIGURE 1a

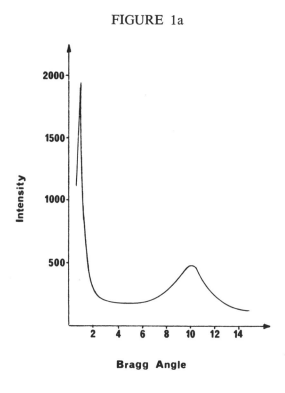

FIGURE 1b

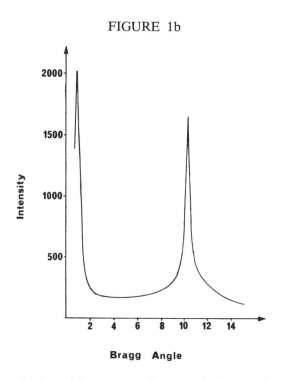

Plots of intensity versus Bragg angle for smectic A (plot a) and smectic B (plot b). Modifications of n-amyl-4-n-dodecyloxybenzylidene-p-aminocinnamate.[7] With permission.

shows that the intermolecular arrangement in the smectic C modification has more order than in smectic A. It can be interpreted that in the transition from smectic A to smectic C on cooling, there is a freezing out of intramolecular conformations, i.e., loss of intramolecular rotation freedom.

From the diffraction peak at small angles it is possible, by use of the Bragg equation, to calculate the strata separation or thickness of the strata of the smectic A, B and C modifications. The value obtained does not always correspond to the length of the molecule but may be somewhat less. Sackmann et al.[25] propose that the shorter distance is due to the non-stretched posture of the hydrocarbon chains in the molecule. Another explanation of the shorter than expected stratum thickness is that the molecules within a stratum may be inclined. For the systems reported by Sackmann and his group, this would require an inclination of the molecules within a stratum of 20–30° with the plane of the stratum. Sackmann and his co-workers don't believe this last explanation is reasonable because optical properties of oriented samples of these compounds show them to be uniaxial; i.e., the long axes of the molecules are perpendicular to the plane of the stratum. The strata spacings were found to be independent of temperature and transition from one smectic texture to another did not alter these spacings. One has to question whether or not the accuracy of measurement at small angles was adequate. A quick check shows that the accuracy has to be better than 10^{-3} degrees to record and measure changes of the order of 1 Å.

The intensity versus angle profiles one gets for the smectic modifications A, B and C are what one would expect from the temperature pattern which is found in the heating and cooling cycle of compounds which exhibit all three textures. Smectic A is the highest temperature form and is next to the nematic or isotropic phase. Smectic C is a lower temperature form than smectic A and smectic B is lower than smectic C. One would expect smectic A to have intramolecular arrangements in strata more like the nematic modification (i.e., random with the long axes of the molecules parallel) and as the temperature is lowered the molecular order should increase, thus requiring smectic C and certainly smectic B to have higher degrees of molecular ordering within strata than found in smectic A.

The smectic D and E modifications have interesting optical and X-ray characteristics. Smectic E has the sharp peak at less than 2° (Bragg angle), just as is found for smectic A, B and C. This peak proves the presence of strata characteristic of smectic textures. In addition to this peak, the diffraction pattern shows three rings at 9.6°, 10.9° and 13.5°. These well-defined rings indicate a high degree of order and pronounced molecular arrangement within strata. The defined order in smectic E suggested the possibility of a two-dimensional structure. Sackmann[25] found that smectic E behaves in a rather unexpected fashion in a magnetic field. In his studies he heated the sample to the isotropic liquid and then cooled it while under a magnetic field (vertical to the X-ray beam) to the test temperature (i.e., smectic E modification). The X-ray pattern of this system shows well-defined strata which proves that the system was completely oriented.

Sackmann and his research team found that smectic D textures behave like isotropic liquids in a polarizing microscope. X-ray patterns show a very weak and blurred ring at a Bragg angle of 10° and the inner ring which was so sharp for the Smectic A, B, C and E modifications degenerated to six points. This structure is comparable to the cubic packing found in lyotropic systems.[31]

Vainshtein, Chistyakov and their co-workers[32-37] have studied the structure of nematic liquid crystals under external fields. In a few paragraphs we will outline the results of several recent publications that have come from their laboratories. The diffuse rings which are characteristic of X-ray patterns of *p*-azoxyanisole in the absence of an electric field are transformed into arcs under a field of 50 v/cm or higher. With a field of 50 v/cm only an orientation effect manifests itself, the long axes of the molecules being oriented normal to the field. With a field of 100 v/cm material flow takes place and this flow partially compensates the orientation of the field. As the field strength goes to 200 v/cm and higher, the orientation from the flow pattern begins to prevail over the orientation by the field. X-ray patterns show orientation of the long axes of the molecules parallel with the direction of the electric field. The schematic representation[36] of the X-ray pattern of *p*-azoxyanisole oriented by a flow arising from an electric field is shown in Figure 2. The arcs and rings are identified as equatorial (E) or meridianal (M). Data relating diffraction angle, θ, and interplanar spacings, d, of *p*-azoxyanisole in an electric field strong enough to generate a flow are recorded in Table 4.

The researchers[32-37] who did this work

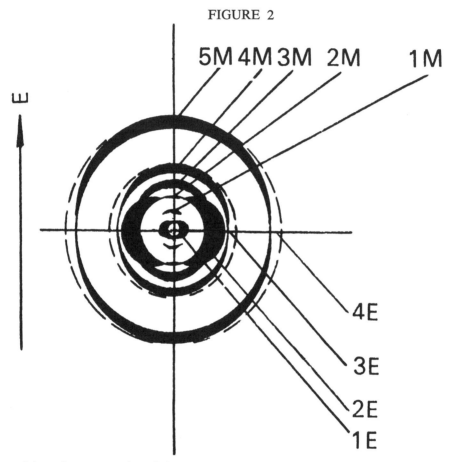

FIGURE 2

Schematic representation of the x-ray pattern of *p*-azoxyanisole in an electric field.[36] With permission.

TABLE 4

Diffraction Angle and Interplanar Distance in *p*-Azoxyanisole in an Electric Field

Radiation	Cu K_α		Mo K_α		Cu K_α		Mo K_α		
Reflection	1E	2E	3E	4E	1M	2M	3M	4M	5M
θ, degree	4°10′	9°33′	10°30′	17°12′	6°	8°33′	14°15′	10°30′	17°21′
d, Å	10.73	4.64	1.96	1.23	7.36	5.18	3.28	1.96	1.23

point out that the strongest equatorial reflections 1E and 2E are due to intermolecular interference and as the field strength increases, the reflections 1E and 2E intensify. This comes about because the more vigorous motion of the liquid crystal orients the molecules more perfectly. As 1E and 2E reflections increase in intensity, the 2M reflection decreases; this strongly suggests that this reflection is also, at least in part, due to intermolecular interferences. The conclusion reached by the authors is that the reflections 1E, 2E and 2M are due to the intermolecular interferences, whereas the reflections 1M, 3M, 4M, 5M, 3E and 4E are due to intramolecular interference.

A magnetic field of 12,000 G on *p*-azoxyanisole resulted in an X-ray pattern[36] which showed more detail than found in the same compound in an electric field. As with the electric field, arcs replaced the diffuse rings which are present in the absence of a field.

We shall make reference to one other X-ray study of a nematic liquid. Gulrich and Brown[15] compared experimental intensities obtained from an X-ray pattern of 4-methoxybenzylidene-4'-cyanoaniline in the liquid crystalline state with calculated intensities of a number of models of molecular packings. A molecular packing which seems to fit the data reasonably well is the "herringbone" packing observed in many organic crystals. The authors point out that other molecular packings are possible. Although this technique of comparing data obtained from X-ray patterns of liquid crystals with models of molecular packings shows promise in helping us to better understand the structure of liquid crystals, it has had little application to date.

IV. THEORIES OF LIQUID CRYSTALLINE STRUCTURES

The three basic types of thermotropic liquid crystals, nematic, cholesteric, and smectic, have one structural property in common. The molecules tend to assume approximate parallel alignment over regions large compared to molecular dimensions. Although this fact has been recognized for a very long time[38] there has been confusion as to the types of long-range order which result from parallel orientation. The reason for confusion is not difficult to perceive. Any realistic theory of liquid crystal structure has to account for the remarkably diverse and subtle changes in physical and optical properties brought about by the effects of surfaces and weak external forces. Therefore, one must be consistently on guard against a superficial or cavalier approach to this subject.

A. The Swarm Theory of Nematic Liquid Crystals

The swarm theory proposed originally by Bose[39] and promoted vigorously by Ornstein[40,41] and Ornstein and Kast[42] has been used extensively in the past to explain many observations such as light scattering and the effect of external fields. The history of this subject is discussed by Brown and Shaw.[3]

According to this theory the interior of a nematic liquid, in the absence of external forces other than orienting surfaces, is composed of clusters of molecules with definite boundaries. A cluster or swarm contains 10^4–10^6 molecules[42] arranged approximately parallel so that a unique axis is determined for the swarm. The swarm axes are assumed to maintain a random thermal arrangement in the liquid crystal. The swarm boundaries are not fixed but can change according to molecular diffusion and Brownian motion. This latter process of change in swarm morphology with time, however, has never been described in detail.

The appeal of the swarm theory is related to the explanation of the effects of electric and magnetic fields which, due to the large net moments, tend to align the swarms in the field direction. If one assumes that the orientation of swarms is subject to thermal fluctuation, one may easily employ a statistical theory similar to that of Langevên to determine such things as saturation effects, size and moment of inertia, dipole moment, magnetic moment, etc. Values for these properties have been reported by many investigators.[3] Also, the existence of swarms of a critical size in the transition region is a natural extension of the theory of phase fluctuations of Ornstein and Zernike.[41]

The swarm concept has been severely criticized, originally by Zocher,[43 44] who explained the physical properties of nematic liquids in terms of an extension of continuum mechanics now referred to as the continuum theory of liquid crystals. The conflict resulting from the interpretation of the swarm theory and the continuum theory has given rise to much of the confusion in liquid crystals mentioned previously. We feel that misunderstanding is so widespread on this point that it is worth a few comments of a general nature. It is now well known that long-range order is the main defining feature of a mesophase. A nematic liquid, for example, is oriented throughout the bulk of the sample even in the absence of external forces or fields. The swarm theory cannot explain this fact. According to the swarm theory one might expect orientation of swarms near surfaces in the absence of external fields, but the swarm directions would become random in the interior of the sample due to thermal fluctuations. The objection to the swarm theory was pointed out in the pioneering work of Zocher and was demonstrated experimentally by Fredericksz and Zolina.[85] On the other hand, the continuum theory is well suited to the treatment of anisotropic liquids because, as we shall see, weak elastic moduli, which are masked in solids, become prominent in liquid crystals and lead to a clear interpretation of their physical properties including the long-range order. The real objection to the swarm theory is not the fact that it is in some way opposed to the continuum theory. It is simply that swarms, as defined originally in nematic liquids, do not exist. Much of the work to be reviewed, in particular that of the past few years, strongly suggests that the statement above is indeed true.

B. The Continuum Theory

The continuum theory of liquid crystals as proposed by Zocher[44] and Oseen[45] is really an important generalization of elasticity theory to include anisotropic liquids. A liquid crystal is assumed to be liquid in the sense that a shear stress is not opposed by permanent forces. However, a new type of deformation not involving a change in distance between points is defined. This deformation involves a change in direction of a vector describing the direction of parallel orientation. It has been called a curvature "strain" and can be induced by a torque "stress." A particularly clear formulation of this theory has been given by Frank.[46] The theory has been generalized recently, particularly in the important series of papers by Ericksen.[47-50] We will outline the theory essentially as found in Frank's paper.

It is proposed that the orientation direction in a liquid crystal can vary slowly from point to point (except at singularities) and is determined by the orienting influence of the container walls and by external forces and fields. It is important to emphasize the orientation forces imposed by surfaces. Indeed, these boundary conditions are so important that we can say that the thermodynamic state of the liquid crystals is not defined unless the boundary conditions are specified. For example, the direction of orientation (shown by lines) for a nematic liquid crystal with parallel orientation at the surface and a magnetic field perpendicular to the surface is shown in Figure 3. In some cases (e.g., cholesteric liquids) a curvature in the direction of parallel orientation may exist in the absence of external forces.

It should finally be noted that the curvature elastic moduli to be described should formally be included in the elastic theory of solids.

FIGURE 3

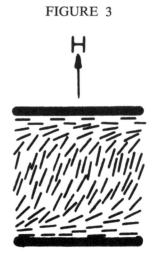

Schematic representation of a nematic liquid crystal with parallel orientation at the surface and a magnetic field perpendicular to the surface.[6]

Frank has emphasized this but has shown that they can be ignored in considering deformations of solids of macroscopic dimension because of the much greater magnitude $\sim 10^{16}$ of the ordinary elastic constants.

C. Curvature Strains in Liquid Crystals

The significant quantity to be considered in the elastic theory of liquid crystals is the tensor gradient $\nabla \mathbf{N}$ of a unit vector $\mathbf{N}$ called the director. This vector is tangential at each point in the liquid to lines tracing out the direction of parallel orientation. In the hydrostatic theory the vector field of the director, including its values on surface boundaries, completely defines the state of deformation in a homogeneous region. The matrix elements of the tensor gradient play the role of the strains in linear elasticity theory. If we choose a coordinate system in which the z axis is along $\mathbf{N}$, $\nabla \mathbf{N}$ has the form:

$$\begin{pmatrix} \frac{\partial N_x}{\partial x} & \frac{\partial N_x}{\partial y} & \frac{\partial N_x}{\partial z} \\ \frac{\partial N_y}{\partial x} & \frac{\partial N_y}{\partial y} & \frac{\partial N_y}{\partial z} \\ 0 & 0 & 0 \end{pmatrix} \quad \begin{array}{ll} s_1 = \frac{\partial N_x}{\partial x}; & s_2 = \frac{\partial N_y}{\partial y} \quad \text{Splay} \\ t_1 = +\frac{\partial N_y}{\partial x}; & t_2 = -\frac{\partial N_x}{\partial y} \quad \text{Twist} \\ b_1 = \frac{\partial N_x}{\partial z}; & b_2 = \frac{\partial N_y}{\partial z} \quad \text{Bend} \end{array}$$

where (s, t, b) are in Frank's notation. The free energy of the liquid crystal is expressed as the volume integral of a free energy density U:

$$G = \int_V U dV, \quad (1)$$

and we assume that U can be obtained by an expansion in powers of the direct product of the gradient tensor $\nabla \mathbf{N}$.[53] If we retain only second order terms, we can formally express U as*

$$U = \text{Tr}(\underset{\sim}{\mathbf{K}})(\underset{\sim\sim}{\nabla \mathbf{N}} - \underset{\sim\sim}{\nabla \mathbf{N}^\circ})^2, \quad (2)$$

where $\mathbf{K}$ is a 9×9 matrix of proper derivatives of U defining the "elastic" constants, and $\nabla \mathbf{N}^\circ$ is a constant matrix which takes account of the fact that the state of lowest energy may not be that for which $\nabla \mathbf{N} = 0$. That is, a state of uniform curvature may be the state of lowest energy (e.g., cholesteric liquids). The inherent symmetry (lack of order) of liquid crystals can be used to obtain a tremendous simplification of the matrix $\mathbf{K}$. Various symmetry conditions and constraints on $\mathbf{N}$ have been discussed by Frank[46] and more recently by Ericksen.[47,49,50] These lead to descriptions of the types of orientation patterns which might occur in nematic, cholesteric and smectic liquid crystals. We will first consider nematic and cholesteric liquid crystals because they are very closely related. The hydrostatic theory of smectic liquid crystals requires special considerations and will be discussed separately.

D. Nematic and Cholesteric Structures

An isotropic liquid is characterized by complete lack of order with respect to translation and orientation. That is, it has the symmetry of the full group $O_3 X T_3$. In a nematic or cholesteric liquid crystal the translational symmetry is retained but the rotational symmetry is broken from O_3 to O_2. That is, there is cylindrical rather than spherical orientational symmetry. Frank[46] uses the O_2 symmetry of U to reduce the number of non-zero constants in (2) from 42 to 7. The general expression for the free energy in Frank's notation is

$$2U = k_{11}(\underset{\sim}{\nabla} \cdot \underset{\sim}{\mathbf{N}} - s_0)^2 + k_{22}(\underset{\sim}{\mathbf{N}} \cdot \underset{\sim}{\nabla} X \underset{\sim}{\mathbf{N}} + t_0)^2 + k_{33}[(\underset{\sim}{\mathbf{N}} \cdot \underset{\sim}{\nabla})\underset{\sim}{\mathbf{N}} \cdot (\underset{\sim}{\mathbf{N}} \cdot \underset{\sim}{\nabla})\underset{\sim}{\mathbf{N}}]$$
$$- k_{12}(\underset{\sim}{\nabla} \cdot \underset{\sim}{\mathbf{N}})(\underset{\sim}{\mathbf{N}} \cdot \underset{\sim}{\nabla} X \underset{\sim}{\mathbf{N}}) - (k_{22} + k_{24})[(\underset{\sim}{\nabla} \cdot \underset{\sim}{\mathbf{N}})^2 - \text{Tr}(\underset{\sim}{\nabla} \underset{\sim}{\mathbf{N}})^2]. \quad (3)$$

---

* In the equations the symbol $\sim$ appearing under a letter will indicate a vector.

In this expression s_o and t_o correspond to an equilibrium configuration with a uniform splay or twist and are directly proportional to the constants which survive in $\nabla \mathbf{N}^\circ$. In a nematic liquid both s_o and t_o are zero or, as stated by Ericksen, $\nabla \mathbf{N} = 0$ is the state of lowest energy in the absence of an applied field or torque. Also Ericksen[47] has shown that k_{24} can be neglected under certain conditions. The vanishing of s_o and also k_{12} is a result of the assumed lack of polarity in a nematic liquid. That is, U is assumed to be invarient under $\mathbf{N} = -\mathbf{N}$. Molecular dipole moments are expected to be distributed randomly with respect to $\mathbf{N}$, although this point is still disputed.* All of these conditions lead to the result that for nematic liquid crystals,

$$2U = k_{11}(\nabla \cdot \mathbf{N})^2 + k_{22}(\mathbf{N} \cdot \nabla \times \mathbf{N})^2 + k_{33}[(\mathbf{N} \cdot \nabla \mathbf{N}) \cdot (\mathbf{N} \cdot \nabla)\mathbf{N}] \quad (4)$$

The constants for splays (k_{11}), twist (k_{22}) and bend (k_{33}) are the same as those defined by Zocher[43] in his continuum theory of nematic liquid crystals. It is important to note that, in principle, these three moduli can be obtained from direct experiments. Saupe[51] has found approximate values based on the study of magnetic deformations. Their magnitude is of the order of 10^{-6} dynes but they are not equal and they show a large temperature dependence.

In a cholesteric liquid crystal one additional term, t_o, is added to the energy expression giving

$$2U = k_{11}(\nabla \cdot \mathbf{N})^2 + k_{22}(\mathbf{N} \cdot \nabla \times \mathbf{N} + t_o)^2 + k_{33}[(\mathbf{N} \cdot \nabla)\mathbf{N} \cdot (\mathbf{N} \cdot \nabla)\mathbf{N}]. \quad (5)$$

In this case the state of lowest energy has a finite twist. The spiral twist can be explained on a molecular basis by the fact that all known cholesteric materials consist of molecules which are optically active, that is, they do not have a plane of symmetry. Hence, the cholesteric structure is the result of enantiomorphy

---

* See later discussion of ferroelectric behavior.

although there is still no physical polarity. The vector $\mathbf{N}$ is uniform (nematic) in each of a family of parallel planes and twists continuously about the normal to these planes. This model of the cholesteric structure is identical to that used by de Vries[52] in his theory of the optical properties of cholesteric liquid crystals. The planes of the director can also be curved surfaces and probably are so under external forces or near singularities. A general discussion of the orientational patterns to be expected in a nematic or cholesteric liquid has been given by Ericksen.[50] It has been pointed out by de Gennes[53] that spiral structures can be expected, in principle, in a medium in which the molecules are not optically active. In this case the energy density has a double minimum; the two minima equally displaced from $\nabla \mathbf{N} = 0$. In this case one might expect a random growth of right or left handed spiral patterns. These structures, however, have never been observed in liquid crystals.

1. Singularities and Distortions—The magnitude of the deformation constants for nematic and cholesteric liquid crystals is indeed exceedingly small. This point has been emphasized in a recent discussion of Zocher.[54] The smallest perturbations due to impurities, foreign particles, surface inhomogeneities, or weak external forces can have a severe and long-range effect on the structure. Hence, on optical examination of a liquid crystal sample, one rarely sees the idealized unperturbed equilibrium configuration. Some structural perturbations are quite prominent as, for example, the threads or nemas observed in nematics and from which they take their name. Certain types of singularities have been treated as analogues of crystal dislocations by Frank[46] and have been termed by him "disclinations." Frank has pointed out that line singularities are possible in nematic liquid crystals because of the lack of a crystal structure. Point singularities are also possible. Oseen[45] calculated the configuration around disclination lines for a nematic liquid with $k_{11} = k_{33}$. In this case the unit vector $\mathbf{N}$ has components in a plane (x_1, x_2) and its direction in the plane is determined by the angle ϕ. The free energy is minimized when

$$\frac{\partial^2 \phi}{\partial x^2} + \frac{\partial^2 \phi}{\partial y^2} = 0. \tag{6}$$

The singular solutions of this equation are

$$\phi = \frac{n}{2} \tan^{-1} x_2/x_1 + \phi_0; \quad \tan \psi = \frac{x_2}{x_1}. \tag{7}$$

where the singularities occur at $x_2 = x_1 = 0$, with a singular line along x_3. Here n is a positive or negative integer giving rise to different deformation contours approaching the singularity. Some of these, according to Frank, are shown in Figure 4.

FIGURE 4

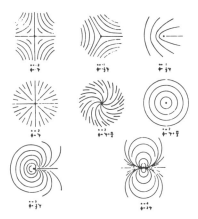

Singuarities resulting in disinclination lines according to Frank. The lines run perpendicular to the plane of the figure.[46]

Most of the patterns shown in the figure have been observed in nematic liquids, but the singularities corresponding to large n are the least likely since the distortion energy increases as n^2.

Another important type of disinclination observed often in cholesteric liquid crystals was first described by Grandjean.[55] The optical patterns are observed only if the cholesteric liquid is confined in a wedge-shaped space provided by slightly tilted glass or mica plates. They appear as bright stripes parallel to lines of equal thickness along the glass surface. These were originally interpreted as planes of discontinuity perpendicular to the viewing surface and have been commonly referred to as Grandjean planes. This description has been criticized in the recent work of Cano.[56] According to Cano the basic factors determining the Grandjean texture in a thin wedge are the orienting influence of the surface and the thickness of the layer, which determine the pitch and the number of half turns of the helix. As shown in Figure 5 a rubbed surface determines the molecular orientation at the two walls. The molecules are shown parallel to the surface and perpendicular to the plane of the figure (x,z plane). It is assumed that an undistorted helix structure is realized only in regions where the gap width is an integral multiple of the half pitch p/2. The half pitch is required because the liquid has no physical polarity; i.e., a rotation about z by π is equivalent to a rotation by 2π. This condition is satisfied at K, K-1, and K-2 in Figure 5. In the regions between these points it is assumed that the number of turns is constant and, in order to satisfy the boundary conditions, the pitch of the helix has to be deformed decreasing with decreasing gap width. The regions of greatest distortion in orientation occur at regular intervals along the x axis midway between the surface planes. The singularities that can occur under these conditions have been discussed by de Gennes.[53] If we move out from a singular point along the y axis, we obtain a line singularity with the lines parallel to regions of equal thickness w. Hence, the Grandjean planes are actually singular lines. This interpretation has been confirmed by the work of Kassubeck and Meier,[57] who have also considered other possible explanations of the Grandjean structure. In particular they have demonstrated, through optical studies of several derivatives of cholesterol, that

FIGURE 5

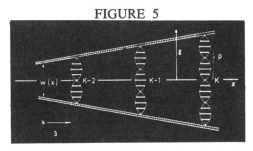

Cross section of a wedge showing the basic assumption that the gap width is equal to an integral multiple of the half pitch.[57]

the gap width is an integral multiple of the half pitch and that the number of turns changes by one half at the optical discontinuities.

Finally, we mention a report of the Orsay Liquid Crystal Group[58] on the observation of a new type of double disinclination line in high pitch cholesteric compounds. These lines presumably give rise to a discontinuous jump of two half pitches. These are characterized by their thickness (they are twice as thick as the commonly observed "first order" lines) and the fact that they break into a zigzag pattern in the presence of a sufficiently high magnetic field. Friedel and Kleman[59] have suggested a new model based on the parallel arrangement of two compensating regular disinclinations of opposite sign.

2. Hydrodynamic Considerations—Some aspects of the hydrodynamic behavior of nematic and cholesteric liquid crystals have been discussed by several investigators.[51,60-64] These are generalizations of the hydrostatic theory discussed previously and are based primarily on the work of Ericksen. This field is only at its beginning due both to the obviously inherent difficulty of the theory and the scarcity of experimental data. The theory has been extended by Leslie,[65-67] Coleman,[68] and Wang.[69] The current status of the theory is discussed in Ericksen.[63] The major emphasis has been on viscosity and the propagation of various types of orientation waves. Earlier references to viscosity measurements in liquid crystals can be found in Gray[8] or Brown and Shaw.[3] One topic of interest is the size effect in viscosity measurements where orientation induced by capillary walls can compete with orientation induced by shear flow. This effect is discussed by Leslie[70] and Ericksen[50] and compared with the experimental results of Porter and Johnson[71] and Porter, Barrall and Johnson.[72]

The propagation of elastic waves in liquid crystals has been considered by Chandrasekhar,[73] Saupe,[74] Ericksen,[50] and de Gennes.[53] The general conclusion is that transverse waves in the low frequency range ($\nu < 10^9$ Hz) are very highly damped. Ericksen has obtained a series of inequalities for the elastic moduli based on general constraints on the energy density. The inequalities are:

$$k_{11} \geq |k_{11} - k_{22} - k_{24}|,$$

$$k_{22} \geq |k_{24}|, \quad k_{33} \geq 0. \quad (8)$$

These inequalities are useful in the consideration of the propagation of orientation waves and guarantee, for example, that the wave velocity of a twist wave $\sqrt{\frac{k_{22}}{c}}$ will be real with a negative damping coefficient.

E. The Smectic Phase

As mentioned previously, the theory of the smectic phase requires separate consideration. This is due to the breaking of the translational symmetry in one dimension. That is, there is one-dimensional translational order. The classification of the smectic phase due originally to Friedel[75] requires that molecules be arranged parallel in layers with the molecular axis normal to the layer. The geometry imposed by the layers, which may be curved surfaces, leads naturally to the explanation of the optically characteristic Dupin cyclides and focal conics.[3] The structure is shown schematically in Figure 6 under smectic A. In terms of the deformation theory smectic A is characterized by comparatively large values of k_{22} and k_{33} so that twist and bend deformations may be neglected, that is, in the energy expression we can set

$$\underset{\text{twist}}{\underset{\sim}{N} \cdot \underset{\sim}{\nabla} \times \underset{\sim}{N} = 0}, \quad \underset{\text{bend}}{(\underset{\sim}{N} \cdot \underset{\sim}{\nabla})\underset{\sim}{N} = 0}, \quad (9)$$

according to Oseen[45] and Frank.[46]

It is now known that some liquid crystal compounds undergo several first order phase transitions between the isotropic and the solid phase, more than can be accounted for on the basis of the classification:

isotropic ⇌ nematic ⇌ smectic ⇌ solid.

A detailed discussion of polymorphism in liquid crystals has been given by Sackmann and Demus.[76] They classify various phases according to texture. According to their classification there is one nematic phase; other transitions occurring at temperatures below the nematic are classified as smectic phases. Five classes are defined, namely, A, B, C, D and E. Some interesting examples of polymorphism in the smectic phase have been reported recently by Fergason et al.[77-79] They have worked with derivatives of compounds such as *bis*(4'-*n*-alkoxybenzal)-1,4-phenylenediamines with as many as eight separate phase transitions. Some of these can be described in terms of the texture classification of Sackmann and Demus. Others cannot be conveniently classified in this scheme. Arora, Fergason and Saupe[80] have also discovered a compound, N,N-(di-*n*-alkoxybenzylidene)-2-chloro-1,4-phenylenediamine, with two phases, both of which have a nematic morphology. The second low temperature phase is assumed to be a smectic C structure in which the molecules are inclined to the normal of the layers. Several possible structural modifications in the smectic phase have been discussed by Saupe.[74] These are shown in Figure 6. In the tilted structure (mentioned above) the continuous rotational axis about C is replaced by a twofold axis, and the liquid crystal becomes biaxial.[79] Other possibilities considered are a twisted smectic, smectic with

FIGURE 6

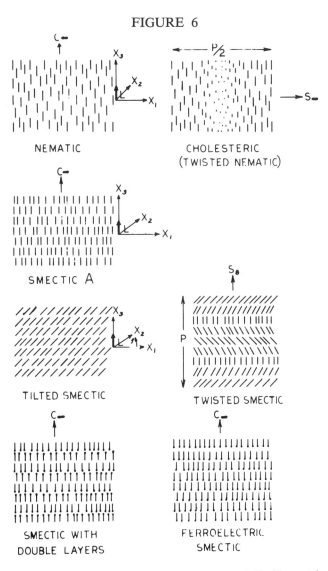

Schematic representation of several types of liquid crystal geometries.[74]

a double layer structure, and a ferro-electric layer structure. Saupe also discusses the modification of the continuum theory imposed on smectic structures which have lost rotational symmetry about the axis normal to the layer. The loss in symmetry leads naturally to a larger number of elastic constants (e.g., nine instead of six in the twisted smectic).

Finally, we mention an interesting material, discussed by Saupe[74] and first described by Lehmann.[81] This is a metastable liquid crystal obtained from a cholesteric compound such as cholesteryl *p-n*-nonylphenyl carbonate by cooling rapidly from the isotropic phase. The liquid is optically isotropic in the undeformed state. It becomes doubly refractive when exposed to slight mechanical stress or surface distortion. Saupe has suggested a structure for this "isotropic" liquid crystal with molecules arranged parallel in large cubic regions, a sort of superlattice of a type discussed by Zocher.[82] This structure (see Figure 7) would have point singularities instead of line singularities.

Although the structures discussed above are highly idealized and tentative, we hope to have emphasized the tremendous subtlety and variety possible in structural modifications of anisotropic liquids.

V. THE EFFECT OF EXTERNAL FORCES AND FIELDS

The structural state of a liquid crystal is very sensitive to external forces as well as boundary conditions at surfaces. The external stimuli which are known to affect the physical and optical properties are static and alternating electric fields, static and rotating magnetic fields, mechanical forces and torques, ultrasonic fields, and even moderately intense electromagnetic waves as have been shown recently by Saupe.[83] The long history of this subject will not be discussed since it can be found in the earlier reviews such as those of Brown and Shaw[3] and Gray.[8] A large amount of work has been reported in this area recently, but it cannot be said that there is general agreement on the interpretation of results. We will discuss the effects of external forces separately for the three different types of thermotropic liquid crystals.

A. Nematic Materials

An effect of primary importance in nematic liquid crystals is the alignment produced by a static electric or magnetic field (see discussion of nmr). These fields are far too weak to produce a torque sufficient to align single molecules at the temperature of the liquid crystal phase. Hence, it has been natural to assume that they act on swarms with large net moments.[42] It was shown long ago by Zocher[44] that it is not necessary to assume the existence of swarms in order to explain field effects. The curvature elastic moduli defined in the continuum theory are very small. Weak external forces can have a pronounced effect on the vector field of the director. There is, in general, a competition between the orientation imposed by the container walls and the orientation imposed by the field. In the case of a magnetic field, the coupling to the director occurs through the anisotropy of the molecular susceptibility associated primarily with the pi electrons of the characteristically aromatic molecules. The energy term which is added to the expression for the free energy density (Equation 4) has the form:

$$U_{mag} = -\tfrac{1}{2}\chi_a (\underset{\sim}{N} \cdot \underset{\sim}{H})^2 , \qquad (10)$$

where $\chi_a = \chi_\parallel - \chi_\perp$ is susceptibility per unit volume. These effects have been discussed by Saupe,[51] de Gennes[53] and Pincus et al.[84] If $\chi_a > 0$, a field applied perpendicular to a containing surface will tend to align the molecular axis parallel to it. This is shown in Figure 8. An important length parameter (ξ) is defined

FIGURE 7

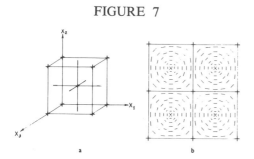

A possible structure of an "isotropic" liquid crystal.[74]

FIGURE 8

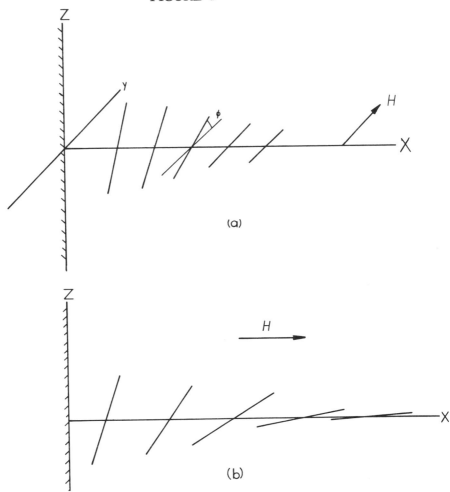

Deformation of a nematic liquid crystal in a homogeneous magnetic field occurring in one coherence length ξ(H). (a) is a pure twist deformation, and (b) is a combination of bend and splay.[53]

which is a measure of the length of the region over which the orientation changes from parallel to perpendicular. The equilibrium configuration in the presence of the field is determined by minimizing the energy density. Under these conditions the parameter ξ is found to be:

$$\xi = \left(\frac{k}{\chi_a}\right)^{1/2} \frac{\pi}{H}, \qquad (11)$$

where k is an average elastic constant for splay and bend. For $k = 10^{-6}$ dynes/cm^2, $\chi_a = 10^{-6}$ cgs and a field of 10^4 G, the coherence length is approximately 1 μ. The concept of the coherence length is useful in the theory of Fréedericksz and Zolina[85] transition and has been discussed by Pincus[84] and de Gennes.[53] Fréedericksz has shown that a nematic liquid between parallel plates and with molecular orientation initially perpendicular to the plates (Figure 9) will assume essentially parallel orientation in a sufficiently large parallel magnetic field. The deformation starts at a critical field H_c which is inversely proportional to the sample thickness d. The analysis of the energy equation under these conditions leads to the result that the transition will not occur unless the coherence length is less than or equal to d/π. It should be noted that there is a large variety of geometries one can choose for the study of deformations in magnetic fields be-

FIGURE 9

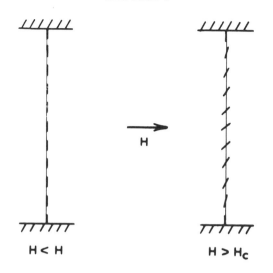

Configuration of a nematic system undergoing the Fréedericksz transition. The coherence length shown is approximately d/π.[53]

cause different types of surfaces are known to orient nematic molecules in many different ways. Even the same liquid crystal can have parallel or perpendicular surface orientation depending on the treatment of the surface.

The field dependence of the magnetic susceptibility of the liquid crystal phase of *p*-azoxyanisole has been studied recently by Massen and Poulis.[86] For magnetic fields below 2000 G the susceptibility is field dependent. A statistical theory based on swarms is used to interpret the data which involves the introduction of an effective mass of a swarm. The saturation effects observed here and similar effects observed with electric fields[87] are not properly interpreted because of the literal use of the swarm concept.

The dielectric anisotropy of nematic liquids and the effects of electric and magnetic fields on dielectric properties have been the subject of many investigations. Since a homogeneously oriented nematic liquid crystal is uniaxial, one can measure two principal dielectric constants $\epsilon_\parallel$ and $\epsilon_\perp$ which are defined with respect to the direction of orientation. The principal dielectric constants for a series of alkoxy derivatives of azo and azoxy benzenes have been reported by Maier and Meier.[88-90] The dispersion over a range of microwave and audio frequencies has been studied by Meier and Saupe[91] and by Axmann et al.[92-94] In these experiments the nematic liquid crystals are oriented by a magnetic field. The normal Debye dispersion is found at microwave frequencies but the azoxy compounds, which have the largest principal value of the dielectric tensor perpendicular to the molecular axis, show an additional dispersion of $\epsilon_\parallel$ at very low frequencies in the radio range. This long-wave dispersion of $\epsilon_\parallel$ has been explained by Meier and Saupe[91] in terms of a relaxation effect based on the intermolecular forces which give rise to the long-range order.

Carr et al.[95-102] have made an extensive study of molecular ordering due to electric and magnetic fields. Their procedure is to study the dielectric loss of a relatively thick sample (1 mm) using microwaves at a fixed frequency. An external magnetic or alternating electric field can be applied simultaneously. Measurements have been made on several different nematic compounds including anisal-*p*-aminoazobenzene with positive dielectric anisotropy and *p*-azoxyanisole with negative dielectric anisotropy. Examples of the dielectric loss curves with simultaneous electric and magnetic fields are shown in Figure 10. The external **E** and **H** are applied in the same direction but, at the frequency of the electric field (300 kHz), the long axes of the molecules are aligned perpendicular to the electric field so that the effects of the two fields actually oppose one another. The ratio E/H corresponding to a dielectric loss (0.515) for equal orientation in the two directions is 1.01. This ratio is essentially constant for values of the magnetic field between 1 and 6 kG. The same results are not obtained for a d.c. electric field. Compounds exhibiting negative dielectric anisotropy such as *p*-azoxyanisole are aligned parallel to a d.c. or low frequency electric field. This behavior has been known for some time,[42] but it is demonstrated most clearly in the dielectric loss curves obtained by Carr. It is apparent that electric field effects are more complicated and not as easy to interpret as magnetic field effects. Alignment in electric fields cannot be explained by the dielectric anisotropy alone. Carr[97] has proposed an explanation based on an anisotropy associated with the electrical conductivity. It should be pointed out that

FIGURE 10

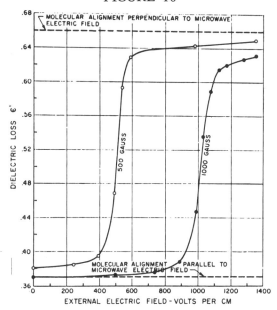

Dielectric loss in *p*-azoxyanisole at a microwave frequency of 24 kMc as a function of an externally applied 370-kc electric field. Individual curves are for various values of a static magnetic field applied parallel to external electric field. The temperature is 132°C. Reproduced with permission from *Ordered Fluids and Liquid Crystals,* Advances in Chemistry Series, No. 63.[101]

FIGURE 11

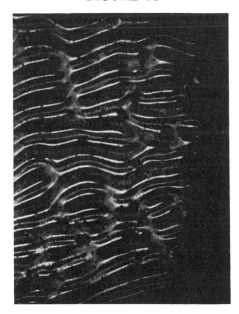

Typical domain pattern. Domain structure in *p*-azoxyanisole. Applied field is 5×10^3 V/cm.[109]

alignment in electric fields is also complicated by the presence of ionizable impurity in the liquid crystal. Nematic liquids such as *p*-azoxyanisole are very difficult to purify completely. Microscopic observation of liquid crystals in d.c. or low frequency a.c. fields shows marked turbulence and streaming, even with so-called very pure samples.

A variety of interesting optical effects has been observed with electric and magnetic fields. The most notable is the formation of domain-like inter-faces in thin films. Williams[103] and, independently, Kapustin and Vistin[104] were the first to observe a domain pattern produced by an electric field on a nematic thin film. These effects have also been discussed by Elliot and Gibson.[105] A typical domain structure is shown in Figure 11 for the nematic phase of *p*-azoxyanisole. This type of mosaic can be produced by d.c. or a.c. electric fields or magnetic fields. The width of the domain pattern can vary considerably with typical values from 0.01 to 10^{-4} cm depending on the film thickness and the frequency of the applied field. A threshold voltage is required to produce the domain pattern which is typically 500–700 V/cm and is dependent on frequency, temperature and the film's thickness. These well-defined optical regions which have been called domains are not to be confused with swarms as defined by Kast. The theory of these electro-optical and magneto-optical phenomena is not well understood. Williams[103,106,107] has proposed a theory based on an analogy with domain formation in ferroelectrics. In this case the direction of the polarization **P** is assumed to be reversed in adjacent domains, minimizing the free energy. This interpretation requires that molecular dipoles have parallel alignment within a given domain; that is, the nematic liquid possesses an intrinsic polarity. Under these conditions, one would expect to observe spontaneous polarization and hysteresis effects as in ferroelectric materials. Experimental attempts to verify this point have been conflicting. No nematic materials are known which possess spontaneous electric polarization. Small hysteresis effects have been reported by Williams and Heilmeier[108,109] and Kapustin and Vistin,[104] but a careful study reported by Kessler et al.[110] indicates that the remanent polarization and non-linearity of the low frequency impedence are due to electrochemical processes rather than to ferroelectricity. Williams has reported further interesting experiments designed to prove that the nematic phase is characterized by intrinsic molecular polarity.[111,112] These are based on the observation of optical rotation in a thin nematic film in a magnetic field. The magnetic field is used to align the liquid crystal and is applied in a direction parallel to the film surface. Light at normal incidence to the film does not show rotation but incident light at an angle to the normal does. The rotation is dependent on the angle of incidence. This effect is not a Faraday rotation because it is independent of field strength (after a threshold value is reached) and occurs for low fields ~2000 G. According to Williams the observed rotation is due to the lack of a center of symmetry in the liquid. There are other possible explanations, however, based on the alignment in the regions containing domain walls which are found under these experimental conditions. The explanation of these small rotations is complicated by the fact that they are not observed at normal incidence. Helfrich[113] has discussed the texture irregularities observed in a magnetic field under the conditions of the Williams experiment. These are called alignment inversion walls and have been related to Bloch walls in a ferromagnetic. The alignment inversion walls are regions where the parallel alignment is reversed by 180°. It is assumed that the molecular orientation is not polar (in contrast to the assumptions of Williams) and that the alignment reversal does not change the physical situation. Several different types of inversion walls are obtained from the analysis of the energy equation of the continuum theory in the presence of a magnetic field. These are shown schematically in Figure 12.

As has been mentioned previously, the effect of electric fields is more difficult to interpret than that of magnetic fields. Some of the unusual optical effects of electric fields have been discussed in relation to domain formation. Another effect called dynamic scattering has been reported recently by Heilmeier et al.[114,115] This has been observed with the nematic compound anisylidene-4-aminophenylacetate. The initially transparent liquid scatters light strongly (appears white) when subjected to d.c. fields of 5×10^3 V/cm. The explanation given by the authors is based on the formation of scattering centers by the transport of ions through the ordered nematic medium. Heilmeier and Zanoni[116] have also used electric fields to orient dichroic dye molecules dissolved in a nematic liquid. Incident polarized light is absorbed when the dye molecules are oriented with the molecular axis parallel to the electric field vector and is transmitted when the molecules are aligned perpendicular to the electric field vector in the presence of the d.c. field. Hence, it is possible to color switch the sample from reddish orange to light yellow by turning on the field. Dichroic studies using polarized infrared light have been reported by Neff, Gulrich and Brown.[87] A d.c. electric field was used to orient the nematic liquid crystals, 4-methoxybenzylidene-4'-cyanoaniline. The dichroic ratio of the $C \equiv N$ bond of this molecule was determined as a function of the applied field. Again, maximum orientation was obtained at a saturation value of $\sim 10^3$ V/cm.

Meyer[117] has outlined a theory which may account for a number of the observed electrooptical effects. This theory is based on the relationship between the Frank curvature strains and the electric polarization in a nematic liquid. Meyer accepts the hypothesis that a nematic liquid crystal is nonpolar (and therefore not ferroelectric) in the absence of external forces. He suggests, however, that either splay or polarization can be induced externally by a mechanical stress or an electric field, respectively. By changing the symmetry of the system from nonpolar to polar, the presence of splay will then induce polarization, or vice versa.

FIGURE 12

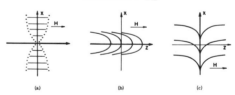

Schematic diagrams of some alignment inversion walls. (a) Twist wall subtended by y and z axes; (b) Splay-bend wall parallel to field; (c) Splay-bend wall vertical to field. The lines represent the direction of orientation.[118]

The effects should be strongest in systems whose molecules possess a large shape polarity as well as a large dipole moment. The theory is developed in analogy with piezoelectric theory in crystals. Since the strains involved in the liquid crystalline piezoelectric effects are curvatures rather than the tensile and shear strains in solids, the geometric aspects of the liquid crystalline effects are entirely different from the effects in ordinary crystals. Since the strains are curvatures, $\nabla \cdot \mathbf{P}$ in which $\mathbf{P}$ is the polarization is seldom zero, so that the strained structures are usually space charged. Also, liquid crystalline curvature electric effects can be expected even though the structure has a center of symmetry. These considerations based on the continuum theory lead to a possible explanation of the sticking together of the spherical droplets formed at the nematic-isotropic transition. The mutual attraction between droplets is explained by the curvature induced space charge. This is shown schematically in Figure 13. Also, the domain patterns observed in weak fields may be explained as a piezoelectric effect which induces curvatures in the structure as shown in the figure. The domain structures are interpreted as alternating regions of bending and splay.

There have been some recent studies of the transport properties in liquid crystals. The thermal conductivity of nematic liquid crystals has been studied by Jules, Picot, and Fredrickson.[118] An electric field, in the direction of heat conduction, produces a higher thermal conductivity. The thermal conductivity is shown to be anisotropic. A theory of heat conduction and dissipation in nematic liquid crystals has been considered by Davison.[119] The electrical conductivity of nematic, smectic, and cholesteric liquid crystals has been studied by Kusabayashi and Labes.[120] Dark- and photo-conductivities of the liquid crystals are shown to be more similar to those of the liquid than the corresponding solids.

Finally, we mention the interesting recent work of Saupe on the effect of a moderately intense electromagnetic field on the structure of the oriented nematic phase.[83] Saupe's experiment demonstrates that a nematic liquid crystal can be deformed by the electric vector of an electromagnetic wave. The material is oriented by a magnetic field and the deformation is observed optically as an increased birefringence of the nematic film. The deformation is produced by a laser with an output of 15 mW at 636 nm.

B. Cholesteric Materials

Studies of the effects of electric and magnetic fields on cholesteric compounds are comparatively recent. The structure of the cholesteric phase is characterized by a uniform twist in the absence of external forces. This twisted structure gives rise to unusual optical effects which can be explained by a selective Bragg reflection determined by the pitch of the helix. Also, the phase is characterized by extremely large optical rotations. These optical effects were first explained by de Vries[52] and have been studied by Fergason et al.[121,122] The wavelength of the reflected light is very temperature sensitive. The temperature variation of the pitch of the helix is not well understood, although an interesting explanation has been given recently by Keating.[123] This is based on the presence of anharmonicity in the forces resisting the relative twist of neighboring planes of molecules and views the macroscopic twist as the rotational analog of thermal expansion in a solid. In order to make this theory

FIGURE 13

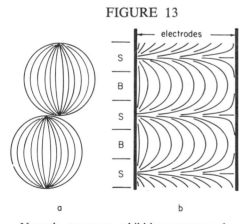

Nematic structures exhibiting curvature-electric effects. The local symmetry axis is parallel to the lines. (a) Nematic droplets attracting one another because of curvature-induced space charges; (b) Cross section of a field induced domain pattern between plane-parallel electrodes, containing alternating regions of splay (S) and bending (B).[117]

compatible with the continuum theory of Frank, one must assume that in the equilibrium configuration there is already a relative twist between molecules in neighboring regions.

In the continuum theory of Frank the cholesteric structure is just a twisted nematic structure as suggested long ago by Friedel.[75] One might then ask whether the cholesteric structure can be untwisted by an electric or magnetic field or by mechanical forces. In the case of a magnetic field, this problem has been considered theoretically by de Gennes.[124] de Gennes considers the application of a magnetic field perpendicular to the axis of the helix (assuming positive diamagnetic anisotropy). His calculations are based on the effect of the field on the elastic free energy in the continuum theory. As the magnetic field is increased, the pitch of the helix increases at first very slowly, and then diverges logarithmically at a critical field $H = H_c$. The relationship between field free pitch and critical field is given by

$$H_c P_o = \pi^2 (k_{22}/\chi)^{1/2}, \qquad (12)$$

where k_{22} is the elastic twist constant and χ is the anisotropic part of the diamagnetic susceptibility. The magnitude of the critical field for a pure cholesteric ester is very high ($H_c \sim 100$ kG). One concludes that a cholesteric to nematic transition would be difficult to observe for pure materials. The actual transition has been observed, however, for a mixture of cholesteric and nematic compounds. This observation has been reported by Durand et al.[126] and compared with de Gennes' theory. For a mixture of p-azoxyanisole and several cholesteryl esters, Durand obtained a critical field for the untwisting of the helix at 9.6 kG. The comparison of the experimentally determined pitch with the curve calculated from de Gennes' theory is shown in Figure 14. The quantitative agreement is very good over the entire range of the magnetic field. An experiment confirming the expected distortion has also been described independently by Meyer.[125] These data provide a means of calculating the elastic constant k_{22} of p-azoxyanisole which compares favorably with that determined by other methods.

The first observations of the untwisting of the cholesteric structure were reported by Sackmann, Meiboom and Snyder.[127] Their results were obtained from studies of nmr spectra of mixtures of active-4,4′-di-*sec.* amylazoxybenzene and 4,4′-di-*n*-hexyloxyazoxybenzene. The significance of these results is the fact that one can apparently obtain the twisted structure with optically active nematic type molecules. A series of such compounds has also been prepared by Leclerq et al.[128] These authors report transition temperatures and heats of transition and find, in some cases, two cholesteric type structures separated by a definite first order transition. This result is difficult to explain on the basis of the accepted concept of the cholesteric structure as a twisted nematic. It prompts us to suggest the need for much further study concerning the proper characterization of a phase, especially in view of the variety of transitions which have been reported recently.

The structure of mixtures of cholesteric esters has been studied by Sackmann et al.[129] The pitch of the helix can be varied by changing the composition of the mixture. The detailed dependence of pitch on composition has been considered by Adams, Haas, and Wysocki.[130] Mixtures of cholesteryl chloride with several cholesteryl esters were studied which exhibit a remarkable sensitivity to composition as shown in Figure 15. As in the case of the temperature dependence, the change in pitch with composition is not well understood

FIGURE 14

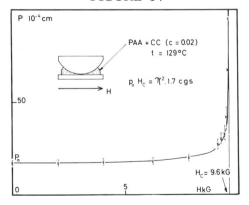

Dependence of pitch P on the field strength H. P_o is the pitch in the absence of the field and H_c is the critical field strength. The theoretical curve is from de Gennes.[124] The experimental data are from Durand et al.[126]

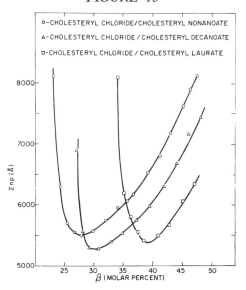

FIGURE 15

Pitch versus chemical composition for three different mixtures of cholesteryl chloride esters.[130]

and no suitable theory has been proposed to explain it.

Adams et al.[131] have also studied optical scattering from cholesteric films subjected to mechanical disturbance by dragging a cover slide over the surface of the liquid crystals. The scattering characteristics of the undisturbed and disturbed materials are quite different. They have shown that the mechanical disturbance causes a reorientation of the optic axis from a direction initially in the plane of the supporting surface to a direction perpendicular to the plane.

Optical studies of the effect of electric fields on cholesteric films have also been reported.[132-134] In low frequency a.c. fields an increase in the intensity of light selectively scattered from the plane texture is found. A variety of complicated optical features such as spots and Maltese crosses is observed, but they are not well understood. The domain patterns characteristic of nematics are not observed. Of particular note is the recent report of Wysocki et al.[135,136] of a phase change in cholesteric liquid crystals subjected to a strong d.c. field. At the transition point which occurs in a field of $\sim 10^5$ V/cm the material changes from optically negative birefringence to positive birefringence. This has been interpreted as a transition from the cholesteric to nematic phase caused again by untwisting of the cholesteric structure.

C. Smectic Materials

There has been no extensive work on the effects of electric or magnetic fields on the structure of the smectic phase. In general, weak fields do not tend to order the smectic phase as they do the nematic or cholesteric phase. This is not surprising due to the greater magnitude of the elastic deformation constants associated with the one-dimensional crystalline order in these materials. It is known that an ordered smectic phase can be obtained by cooling from the nematic phase in the presence of a magnetic field.

The formation of a domain pattern in a strong electric field has been reported by Vistin and Kapustin.[137] The compound, 4-n-heptyloxybenzoic acid, has a domain pattern typical of that observed in nematics in a 12 Hz electric field of 7 kV/cm. The threshold value of E increases with increasing frequency.

VI. THERMODYNAMICS OF THERMOTROPIC LIQUID CRYSTALS

In comparison with the vast amount of information on the thermodynamic properties of solids and isotropic liquids, the number of significant thermodynamic studies on liquid crystals is extremely sparse. There is a real need for equation of state data and phase diagram studies on single component thermotropic liquid crystals.

The anistropy in the physical properties of liquid crystals is a manifestation of long-range orientational order. One object of the thermodynamic studies is to relate variables such as volume, temperature and free energy to suitably defined order parameters as in the case of order-disorder phenomena in solids. First order phase transitions in liquid crystals provide interesting examples of cooperative phenomena, and it is surprising to note that the recent monographs on order-disorder transitions make no mention of them.

A. Statistical Theory of Phase Transitions

Before considering the existing statistical theories in detail, we wish to discuss a general classification scheme based on symmetry considerations. Brout[138] has classified phase transitions in a very general way according to a group of continuous symmetry in the diordered phase which is broken in the ordered phase. When the symmetry of the continuous group is broken, a set of collective modes arises in the condensed phase. The zero frequency limit of the continuous spectrum of the collective modes corresponds to the macroscopic motion which is invariant under the continuous group. Brout gives several examples of transitions classified in this way in tabular form. To this list we can add the various types of liquid crystal phase transitions. In the isotropic liquid we have, in addition to the continuous translational symmetry of the liquid, T_3, the continuous group of displacements about a fixed point or the rotation group O_3. In the transition of the isotropic liquid to the nematic or cholesteric phase, the rotational symmetry is broken from O_3 to O_2.

$$\begin{array}{ccc} T_3 \times O_3 & \longleftrightarrow & T_3 \times O_2 \\ \text{isotropic} & & \text{nematic or cholesteric} \\ \text{liquid} & & \text{liquid crystal} \end{array}$$

The collective modes which arise in the nematic phase can be described as orientation waves analogous to spin waves in a ferromagnet. In the transition from the nematic phase to smectic A, one degree of translational symmetry is broken:

$$\begin{array}{ccc} T_3 \times O_2 & \longleftrightarrow & T_2 \times O_2 \\ \text{nematic} & & \text{smectic A} \\ \text{phase} & & \end{array}$$

The collective excitations in this case would be one-dimensional longitudinal phonons. Similar considerations can be used to describe transitions to the other types of smectic phase (see Section III). The relevance of this classification scheme has been discussed in detail by Brout. In each case of broken continuous symmetry one can define response coordinates and response functions which can be related by the methods of statistical mechanics. The response coordinate undergoes infinite fluctuations at the transition temperature and, in the case of liquid crystals, can be related to the degree of order S.

Liquid crystalline phase transitions are typically first order with latent heats ~ 0.1–0.7 kcal/mole (nematics and cholesterics) and 1–3 kcal/mole (smectics). The volume changes for the isotropic-nematic transition are quite small, generally about one order of magnitude less than for comparable liquid-solid transitions. A statistical theory of the nematic phase has been given by Maier and Saupe.[139,140] The statistical model is based on an average internal field of the Weiss type. The nature of the assumptions concerning the intermolecular forces giving rise to liquid crystallinity is a very important feature of this theory. Maier and Saupe give convincing evidence that the forces predominantly responsible for parallel orientation in the nematic phase are the dipole-dipole dispersion forces which are highly angularly dependent because of the anisotropy of the optical transition moments of the elongated aromatic molecules. For example, they neglect the attractive forces due to permanent dipoles because these would lead to ferro-electric order and spontaneous polarization which have not been observed in the nematic phase. Also, liquid crystals have been found which do not have a permanent dipole moment.[141] Maier and Saupe derive an expression for the average energy of the ℓ th molecule oriented at an angle θ to a fixed spatial direction in the presence of the average interaction field of all the other molecules. The average internal field is expressed in terms of the order parameter S (see discussion on magnetic resonance).

$$E_\ell = -\frac{A}{V^2} S(1 - \frac{3}{2} \sin^2\theta_\ell) , \quad (13)$$

where

$$S = (1 - \frac{3}{2} \sin^2\theta)$$

and

$$\frac{A}{V^2} = \sum_{ij} \frac{\delta_{oi}\delta_{oj}}{E_{ij}-E_{oo}} \sum_k \frac{C}{(R_{\ell k})^6} ,$$

is a term independent of orientation containing the volume ($R_{\ell 2}^6$) dependence and the molecular transition moments whose magnitude determines the strength of the dispersion force. The average orientation of the *l*th molecule is obtained from the distribution taken over E_l.

$$\overline{\sin^2\theta_l} = \frac{\int_0^{\pi/2} \sin^3\theta_l \, e^{-\frac{E_l}{kT}} \, d\theta_l}{\int_0^{\pi/2} \sin\theta_l \, e^{-\frac{E_l}{kT}} \, d\theta_l} \quad (14)$$

The self consistency condition

$$\overline{\sin^2\theta_l} = \overline{\sin^2\theta} \quad (15)$$

leads to the allowed thermodynamic states as shown in Figure 16. The Point P_1 with positive slope is a possible state of stable nematic order. The point P_2 corresponds to the stable isotropic phase where $S = 0$. The quantity A/kTV^2 determines the shape of the curves shown and, therefore, the degree of order for the transition. The points P_2 and P_1 are separated for all values of the parameter A/kTV^2 and the transition is first order. The point P_3 on curve (4) represents the maximum value of $\overline{\sin^2\theta}$ or the minimum value of S which is attained on discontinuous transition from the isotropic phase. This occurs for $S = 0.32$. It should be noted that extrapolation of experimentally determined values of S always leads to a limiting value greater than the minimum predicted by the theory of Maier and Saupe (see discussion on magnetic resonance). In addition to the introduction of the parameter S which is a measure of the long-range order and is analogous to the magnetization in the Weiss theory, Maier and Saupe[139] recognize the need for a short-range order parameter, taking account explicitly of nearest neighbor interactions and packing considerations of the elongated molecules. They account for this in a very simple way by assuming that the nematic molecules form small spherical clusters or cybotactic groups which persist into the disordered phase. The combination of this assumption with the expression derived for the free energy change in terms of S leads to the derivation of a reduced equation of state for the order parameter which is proposed to be universally valid. The reduced variable is $TV^2/T_kV_{n,k}^2$ where T_k is the temperature and $V_{n,k}$ is the molar volume at the clearing point. It is also concluded that, at the clearing point, all nematic liquids have approximately the same state of order ($S \approx 0.44$), and this result seems to be verified experimentally in terms of those compounds for which S has been measured as a function of temperature. Saupe[142] has used the thermodynamic expression for the enthalpy in terms of the order parameter S to treat the temperature dependence of the elastic deformation constants of nematic liquids. The configurational change in the enthalpy expression is related to Frank's expression for the deformation energy in terms of the curvature elastic constants. Hence, the temperature dependence of the elastic constants is related directly to the temperature dependence of the degree of order S. Saupe obtains

$$K_l = \frac{S^2}{V^{7/3}} C_l, \quad (16)$$

where S is the degree of order, V is the volume and C_l is a temperature independent constant characteristic of the type of deformation, i.e., twist, splay or bend.

The statistical theory of Maier and Saupe is the first serious attempt to study the theory of liquid crystallinity based on a molecular

FIGURE 16

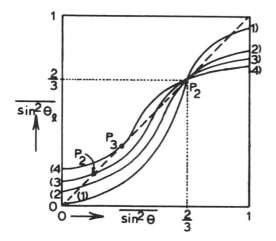

$\sin^2\theta_l$ as a function of $\sin^2\theta$ with A/kTV^2 as parameter. Curve (1) $A/kTV^2 > 5$; (2) $A/kTV^2 = 5$; (3) $4.4876 > A/kTV^2 < 5$; (4) $A/kTV^2 = 4.4876$.[142]

theory. We should point out, however, that the beginnings of such a theory based on the Weiss field were considered some time ago by Tsvetkov.[143] The advantages and limitations of the internal field model, which amounts to the assumption of interactions of infinitely long range, are well known in statistical mechanics.[144] We would expect that certain thermodynamic properties, such as heat capacity, would not be well determined by a model based on the internal field. This is borne out by the experimental determinations of Arnold.[145] Cotter and Martire[146] have attempted to take account of short-range interactions by treating the nematic isotropic transition of a system of rigid rods in terms of a lattice model. The repulsive forces are accounted for directly in terms of the packing of rods of given length-to-breadth ratio. Intermolecular attractions were accounted for in terms of a segmental interaction energy. The partition function was evaluated by the quasi-chemical approximation method. Two stable phases were found, one corresponding to the completely aligned phase and the other to the isotropic phase. The attractive energy term was necessary in order to obtain transition temperatures for realistic length-to-breadth ratios. The transition temperature in terms of the length-to-breadth ratio (x), the volume fraction of holes v_o, and the segment interaction energy w_{xx} are shown in Figure 17. The dependence of the enthalpy of transition on the segment interaction energy is shown in Figure 18. The model is able to explain the concurrent decreases in transition temperatures and increases in enthalpies of transition which occur in ascending homologous series of nematic compounds.

It is apparent that the statistical treatment of phase transitions and order in liquid crystals based on molecular models is only in the beginning phase. As has been previously mentioned, it is surprising that the methods of statistical mechanics have not been applied more extensively to these systems. Certainly the intermolecular forces are not well understood,

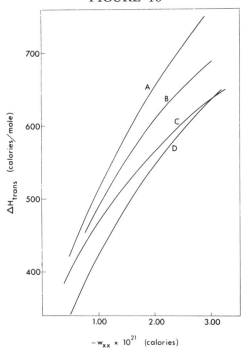

FIGURE 17

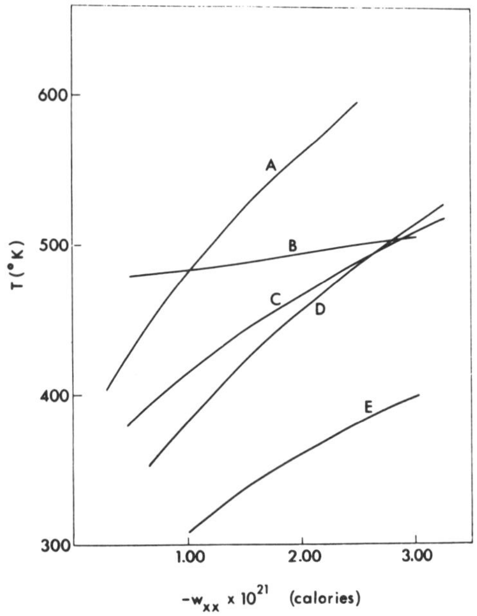

Dependence of the transmission temperature T on the segment interaction energy w_{xx}.[146]
(A) $x = 10$, $v_o = 0.65$ $(w_{xx'} - w_{xx}) = 4.0 \times 10^{-23}$ cal.
(B) $x = 5$, $v_o = 0.15$ $(w_{xx'} - w_{xx}) = 1.5 \times 10^{-23}$ cal.
(C) $x = 7$, $v_o = 0.50$ $(w_{xx'} - w_{xx}) = 4.0 \times 10^{-23}$ cal.
(D) $x = 10$, $v_o = 0.65$ $(w_{xx'} - w_{xx}) = 3.0 \times 10^{-23}$ cal.
(E) $x = 7$, $v_o = 0.55$ $(w_{xx'} - w_{xx}) = 4.0 \times 10^{-23}$ cal.

FIGURE 18

Depedence of the enthalpy of transition ΔH on the segment interaction energy w_{xx}.[146]
(A) $x = 10$, $v_o = 0.65$ $(w_{xx'} - w_{xx}) = 4.0 \times 10^{-23}$ cal.
(B) $x = 7$, $v_o = 0.55$ $(w_{xx'} - w_{xx}) = 4.0 \times 10^{-23}$ cal.
(C) $x = 7$, $v_o = 0.50$ $(w_{xx'} - w_{xx}) = 4.0 \times 10^{-23}$ cal.
(D) $x = 10$, $v_o = 0.65$ $(w_{xx'} - w_{xx}) = 3.0 \times 10^{-23}$ cal.

but theories of the phase transition should be less difficult than, for example, melting where the configurational order in the condensed phase is far more subtle than simple parallel alignment. One is tempted to consider other models, such as the Ising model, which take account explicitly of the short-range attractive forces. Indeed, recent work carried out at the Kent Liquid Crystal Institute shows that the Ising model can be properly adapted to account for the isotropic-nematic phase transition. Part of this work focuses on the equation of state and suggests that there may be a critical temperature at or above which the nematic-isotropic transition becomes second order. There are currently no experimental data available at sufficiently high pressures to confirm or refute this speculation. Finally, we mention other work in the Institute, soon to be published, based on a modification of the Lennard-Jones-Devonshire model of melting. The model leads, in general, to two transitions at which positional and orientational order are lost. By adjustment of the relative energies associated with these transitions a nematic phase becomes apparent. The parameters of the transitions associated with this phase are qualitatively similar to the experimental results on nematic materials.

B. Thermodynamic Measurements

On cooling from the isotropic liquid phase, through the various mesophase transitions, the latent heats and entropy changes are generally quite small compared to the final transition from mesophase to crystalline solid. For example, the transition heats (Kcal/mole) for the phase transitions of 4,4′-di-n-heptyloxyazoxybenzene are

$$\text{isotropic} \xleftrightarrow{0.243} \text{nematic} \xleftrightarrow{0.381} \text{smectic} \xleftrightarrow{9.78} \text{crystalline solid.}$$

Such small changes are difficult to measure accurately, and it is not surprising that there has been widespread disagreement on the precise values for the thermodynamic quantities. Additional problems arise due to difficulties in purification and decomposition at the high temperatures required for most mesophase transitions. Hence, systematic thermodynamic studies have been made on only a few of the many liquid crystal compounds. Recent thermodynamic work has been reviewed and discussed by Porter, Barrall and Johnson.[147,148]

Most of the thermodynamic studies have centered about nematic or cholesteric compounds in homologous series. It has been known for some time that there is a progressive alternation in the clearing point temperatures of a homologous series of nematic compounds with increasing chain length of a substituted alkyl chain as shown in Figure 19. This interesting odd-even regularity has been discussed by Maier and Saupe in connection with the effect of the extended alkyl chain on the dipole-dipole dispersion forces.[140] A similar alternation is shown for a series of cholesteryl n-alkanoates (Figure 20). In the case of the nematic compounds, similar alternation regularities occur in the latent heats and entropies of the mesophase isotropic transition but not in the crystal mesophase transition. Calorimetric measurements using the method of differential thermal analysis and differential scanning calorimetry have been reported recently by Barrall, Porter and Johnson[149-152] and by Ennulat.[20,153]

The most significant thermodynamic studies have been reported in an important series of papers by Arnold et al.[154-161] Arnold uses the method of adiabatic calorimetry for the determination of heat capacities and latent heats. We have summarized the latent heat data of Arnold in Tables 5, 6, and 7. This work can be organized into three categories: (a) a homologous series of nematic compounds consisting of the 4,4′-di-n-alkoxyazoxybenzenes; (b) a series of cholesteryl esters; and (c) a series of compounds with various intermediate smectic phases classified according to the method of Sackmann (see Section III). Arnold[160] has also studied the latent heats of a long series (n = 2 — 15) of cholesteryl esters of saturated fatty acids by the methods of differential thermal analysis. A comparison of his results with those of Barrall, Porter, and Johnson[152] is shown in Figure 21. The rather large discrepancy in these values is an indication of the uncertainty of the DTA method and suggests that caution should be exercised in the interpretation of results based on this method.

We can summarize the results of the thermodynamic studies in terms of some general statements. The entropy changes of nematic-isotropic and cholesteric-isotropic transitions are of the same order of magnitude in agreement with accepted interpretation regarding the similarity of these phases. In homologous series, aside from the small odd-even alternations, there is a general tendency for the transition entropy to increase with increasing molecular weight. A general correlation is observed[148] in which the mesophase-isotropic transition entropy is roughly 2% of the total for all transitions from solid to isotropic liquid. Smectic-isotropic transition entropies are usually much larger, roughly by a factor of ten. For intermediate transitions the smectic B to smectic C is larger than that for smectic C to smectic A.[159] For example, from Table 7 we see that the transition entropy for smectic C to smectic A for di-n-dodecyl 4,4'-azoxy-α-methylcinnamate acid is only 0.069 cal/mole °K.

FIGURE 19

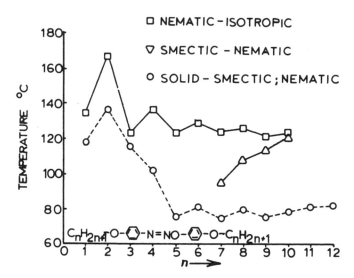

Melting and clearing temperatures for a series of dialkoxyazoxybenzenes. n is the number of carbon atoms in the alkoxy chain.[145]

FIGURE 20

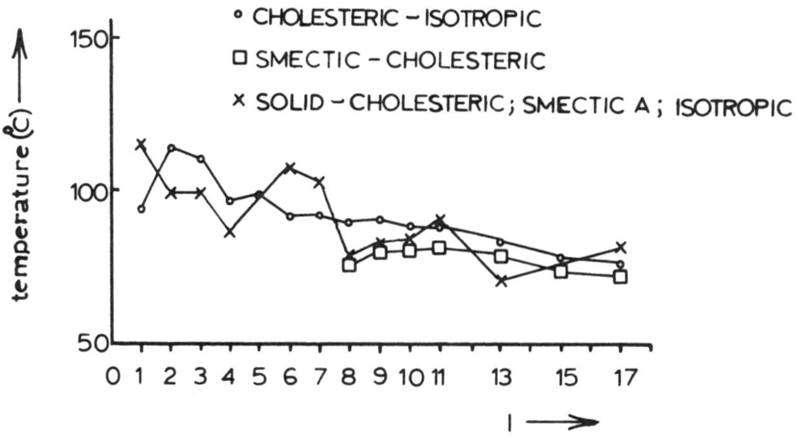

Melting and clearing temperatures for a series of n alkyl cholesteric esters.[109]

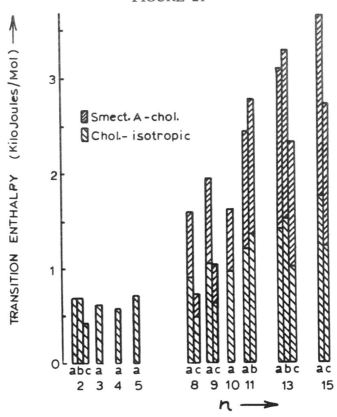

Transition enthalpies for a series of cholesteryl n-alkanoates. (a) DTA values according to Arnold.[160] (b) Values obtained from adiabatic calorimetry according to Arnold.[158] (c) DTA values according to Barrall, Porter, and Johnson.[152]

Arnold has also obtained heat capacity data for the compounds listed in Tables 5, 6, and 7. For example, the heat capacity curve through the clearing point for p-azoxyanisole is shown in Figure 22. The curves are generally unsymmetrical with a gradual increase in slope on the low temperature side of the transition. The heat capacity curve for a compound exhibiting several mesomorphic phases is shown in Figure 23. Arnold[145] has compared his heat capacity data with the statistical theory of Maier and Saupe. In general, the details of the heat capacity curve are not well determined by the theory. One of the serious problems in this area is the lack of data on compressibility and thermal expansion coefficients. p-Azoxyanisole is one of the few compounds for which these data are available.[140] Unfortunately, it is one of the most difficult to work with from the point of view of sample purification and thermal decomposition. Arnold emphasizes the effect of impurity on the shape of the heat capacity curves. Thermal expansion coefficients have been reported recently for some cholesteryl esters.[162] Pretransition effects in the nematic-isotropic and the cholesteric-isotropic transitions are also observed in the thermal expansion studies. It seems to be established that mesophase transitions are more extended than, for example, the melting transition. Torgalkar and Porter[163] have discussed pretransition effects in liquid crystals according to the heterophase fluctuation theory of Frenkel. They have pointed out that anomalies in physical properties near mesophase transitions have been observed in nmr relaxation times,[164] coefficients of expansion,[165] ultrasonic absorption,[166,167] viscosities,[165] gas chromatographic retention times,[150] surface tension,[168] and magnetic flow birefringence.[169]

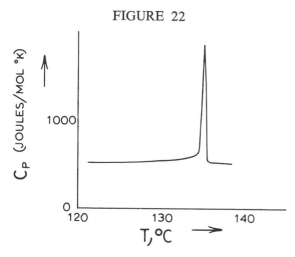

Heat capacity curve for *p*-azoxyanisole according to Arnold.[155]

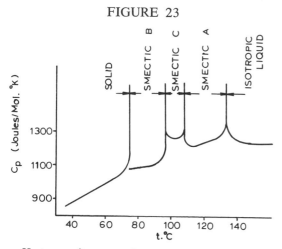

Heat capacity curve for *n*-amyl, 4-*n*-dodecyloxy-benzylidene-4-aminocinnamate.[159]

Pretransition effects on the electro-optical properties of liquid crystals have been considered recently by Tsvetkov and Ryumtsev.[170] Heterophase fluctuations are expected to be most pronounced under conditions where latent heats are very small and there is considerable similarity between the two phases. Mesophase transitions are exactly of this type and provide a fertile field for the study of fluctuation phenomena.

VII. INFRARED AND RAMAN SPECTRA OF LIQUID CRYSTALS

Very few detailed spectroscopic studies of liquid crystals have been reported. Most of the work that has been done has been concerned with the orientation of nematic compounds in electric or magnetic fields. Polarized infrared radiation interacts with a molecular vibrational transition moment according to the expression $(\mathbf{M} \cdot \mathbf{E})^2$. If the transition moment ($\mathbf{M}$) is perpendicular to the field vector $\mathbf{E}$, no light is absorbed. Using polarized infrared radiation, Maier and Englert[171][172] have studied the degree of orientation in nematic liquid crystals oriented by surface forces. Neff, Gulrich and Brown[87] have studied the orientation in a d.c. electric field by infrared dichroic methods. The Raman spectrum of 4-*n*-butyloxybenzoic acid has been studied in a longitudinal and transverse magnetic field by Koller, Lorenzen, and Schwab.[173] They find that certain vibrations become active in a magnetic field and that the activity depends on the orientation of the vibrational polarizability tensor with respect to the electric vector of the scattered light.

In general, aside from the effect of orientation on absorption intensity, the infrared spectra of nematic and cholesteric liquid crystals are almost identical to those of the isotropic liquids. There is no evidence to indicate any pronounced change in intermolecular interactions or short-range order in passing from the isotropic to the nematic phase. Some small frequency shifts have been observed in the infrared spectra of smectic compounds indicating a more pronounced change in short-range intermolecular forces in the smectic phase.

VIII. LIGHT SCATTERING AND SPIN-LATTICE RELAXATION

Nematic liquid crystals strongly scatter light as evidenced by their milky or turbid appearance. In fact, one can easily recognize the nematic phase by this characteristic appearance. When heated beyond the nematic temperature range into the isotropic liquid the sample loses its turbidity and becomes clear. The ability of the liquid crystal to strongly scatter light intrigued several early investigators and was thought to hold the clue to the structure of the liquid crystal phase. Measurements of the angular dependence of the scattered intensity and polarization were made by

TABLE 5

Transition Temperatures, Enthalpies and Entropies for a Series of 4,4'-di-n-alkoxyazoxybenzene
(According to H. Arnold)

RO-⟨phenyl⟩-N=N(→O)-⟨phenyl⟩-OR

Alkyl Group	Transition Temp. °C			ΔH Kcal/mole			ΔS cal/mole °K		
	Solid	Smectic-Nematic	Nematic-Isotropic	Solid	Smectic-Nematic	Nematic-Isotropic	Solid	Smectic-Nematic	Nematic-Isotropic
methyl	118.2		135.5	7.067		0.137	18.06		0.335
ethyl	136.6		167.5	6.422		0.326	15.67		0.740
propyl	115.5		123.6	6.429		0.161	16.54		0.406
butyl	102.0		136.7	5.005		0.247	13.34		0.603
pentyl	75.5		123.2	3.487		0.173	10.01		0.436
hexyl	81.3		129.1	9.892		0.250	27.91		0.621
heptyl	74.4	95.4	124.2	9.780	0.381	0.243	28.14	1.032	0.613
octyl	79.5	107.7	126.1	10.081	0.282	0.348	28.58	0.741	0.860
nonyl	75.5	113.0	121.5	9.123	0.395	0.422	26.25	1.021	1.07
decyl	78.2	120.6	123.4	9.221	0.554	0.752	26.24	1.406	1.90
		Smectic-Isotropic			Smectic-Isotropic			Smectic-Isotropic	
undecyl	80.8	121.4		9.82	2.407		27.75	6.10	
dodecyl	81.7	122.0		10.05	2.861		28.34	7.24	

TABLE 6

Transition Temperatures, Enthalpies, Entropies for a Series of Cholesteryl Esters
(According to H. Arnold)

	Transition Temp. °C			ΔH Kcal/mole			ΔS cal/mole °K		
	Solid	Smectic-Cholesteric	Cholesteric-Isotropic	Solid	Smectic-Cholesteric	Cholesteric-Isotropic	Solid	Smectic-Cholesteric	Cholesteric-Isotropic
propionate	99.6		114.1	5.21		0.16	13.98		0.413
laurate	91.3	81.4	88.6	7.91	0.340	0.325	21.70	0.959	0.898
myristate	71.0	79.1	84.6	11.35	0.425	0.361	32.98	1.21	1.01
ethyl-carbonate	83.7		105.7	5.04		0.17	14.13		0.449

Chatelain.[174 175] The general features of his results were that the scattered intensity was very strong at low angles and that liquid crystals strongly depolarize light. That is, if a linearly polarized beam of light is incident upon the liquid crystal, the scattered light will contain components of polarization perpendicular to that of the incident beam. It was believed by some that this scattering was due to large particles or swarms. Others pictured a continuous model undergoing long wavelength order fluctuations. Recent calculations[176] of fluctuation phenomena in liquid crystals appear to confirm this latter picture. de Gennes and the Orsay Liquid Crystal Group have calculated long wavelength order fluctuations. Not to be confused with propagating modes, their frequencies are imaginary; that is, when the preferred direction of orientation of the molecules is thermally disturbed, it returns to equilibrium without any oscillations. Based upon these order fluctuations, de Gennes calculated the differential scattering cross section and found that he could explain Chatelain's earlier measurements[53] to include the functional dependence of the scattered intensity with angle.[177] Further justification of this theoretical approach can be seen in nuclear spin-lattice relaxation measurements.[178] In these studies one measures a spin-lattice relaxation time, T_1, which is a measure of the rate at which energy is exchanged between the nuclear spins and the lattice. The quantity T_1 depends upon how the liquid crystal molecules are individually or collectively modulated by the thermal action. A simple calculation of T_1 based upon the order fluctuations shows that this quantity should vary as $\omega^{1/2}$ where ω is the Larmor precession frequency. This frequency dependence is unusual for liquids in general but is, in fact, what is observed in nematic as well as cholesteric and smectic liquid crystals.

The theoretical treatment of de Gennes[176] and the Orsay group describes fluctuations in the director $\mathbf{N}(\mathbf{r})$ about some fixed director established by the unit vector $\mathbf{N}_0$. For small deviations about $\mathbf{N}_0$

$$\mathbf{N}(\mathbf{r}) = \mathbf{N}_0 + \delta\mathbf{N}(\mathbf{r}) \qquad (17)$$

at the point $\mathbf{r}$ in the liquid crystal. The physical meaning of the vector $\mathbf{N}$ is the same as one would attach to the director in Frank's continuum theory. To attach the vector $\mathbf{N}(\mathbf{r})$ directly to the long axis of a molecule would not necessarily be in accord with the theory. This is particularly true when $\mathbf{N}(\mathbf{r})$ is considered to deviate only slightly from $\mathbf{N}_0$ as in Equation 17. The long axis of the molecule itself may swing 20° to 40° from its preferred direction. This can be seen quite simply by considering the order parameter $S = \frac{1}{2} \langle 3\cos^2\theta - 1 \rangle$ which is known to vary from 0.7 to 0.3. One should more appropriately associate $\mathbf{N}_0$ to the preferred or average direction

TABLE 7

Transition Temperatures, Enthalpies, and Entropies of a Selected Group of Compounds

n = 10   decyl
         di-n-   ester of 4,4'-azoxy-α-methylcinnamic acid
n = 12   dodecyl

$$C_nH_{2n+1}OOC-C(CH_3)=CH-\langle\bigcirc\rangle-N=N(\to O)-\langle\bigcirc\rangle-CH=C(CH_3)-COOC_nH_{2n+1}$$

	Transition Temp. °C			ΔH Kcal/mole			ΔS cal/mole °K		
	Solid	Smectic C-Smectic A	Smectic A-Isotropic	Solid	Smectic C-Smectic A	Smectic A-Isotropic	Solid	Smectic C-Smectic A	Smectic A-Isotropic
n = 10	65.8	73	88.1	10.875	0.024	1.768	32.08	0.069	4.893
n = 12	78.6	82.5	87.5	17.973	0.024	2.103	51.09	0.069	5.830

$$C_{18}H_{37}-O-\langle\bigcirc\rangle-N=NO-\langle\bigcirc\rangle-O-C_{18}H_{37}$$

Transition Temp. °C			ΔH Kcal/mole			ΔS cal/mole °K		
Solid	Smectic B-Smectic C	Smectic C-Isotropic	Solid	Smectic B-Smectic C	Smectic C-Isotropic	Solid	Smectic B-Smectic C	Smectic C-Isotropic
94.1	99.0	115.3	18.88	2.52	5.40	51.40	6.77	13.89

$$C_{12}H_{25}-O-\langle\bigcirc\rangle-CH=N-\langle\bigcirc\rangle-CH=CH-COOC_5H_{11}$$

Transition Temp. °C				ΔH Kcal/mole				ΔS cal/mole °K			
Solid	Smectic B-Smectic C	Smectic C-Smectic A	Smectic A-Isotropic	Solid	Smectic B-Smectic C	Smectic C-Smectic A	Smectic A-Isotropic	Solid	Smectic B-Smectic C	Smectic C-Smectic A	Smectic A-Isotropic
74.6	96.3	107.3	133.1	6.692	1.309	0.411	2.017	19.24	3.543	1.080	4.965

TABLE 7 (Continued)

diethyl 4,4'-azoxybenzenedicarboxylate

$C_2H_5OOC-\langle\underline{\ }\rangle-N=NO-\langle\underline{\ }\rangle-COOC_2H_5$

Transition Temp. °C		ΔH Kcal/mole		ΔS cal/mole °K	
Solid	Smectic A-Isotropic	Solid	Smectic A-Isotropic	Solid	Smectic A-Isotropic
114	122.6	4.737	1.211	12.23	3.06

ethyl 4-ethoxybenzylidene-4'-aminocinnamate

$C_2H_5-O-\langle\underline{\ }\rangle-CH=N-\langle\underline{\ }\rangle-CH=CH-COOC_2H_5$

Transition Temp. °C				ΔH Kcal/mole				ΔS cal/mole °K			
Solid	Smectic B-Smectic A	Smectic A-Nematic	Nematic-Isotropic	Solid	Smectic B-Smectic A	Smectic A-Nematic	Nematic-Isotropic	Solid	Smectic B-Smectic A	Smectic A-Nematic	Nematic-Isotropic
81.7	118.5	156.5	159	6.530	0.502	1.219	0.120	18.40	1.28	2.84	0.276

of the individual molecules. In this case we would be considering fluctuations of the preferred direction $N(r)$ about the average direction N_o while the molecules themselves may be executing additional short-range motion about the director $N(r)$. Just exactly how the axis of the molecules is fluctuating about $N(r)$ or what the connection is between the short-range order and fluctuations described by de Gennes is yet unclear. This particular uncertainty becomes a point of concern when considering spin-lattice relaxation which will be discussed later.

In their calculations, the Orsay group arrives at two eigenfrequencies (one fast and one slow) for the fluctuation modes. The general expressions for these frequencies contain the six Leslie friction coefficients[65] as well as Frank deformation constants.[46] Although the values of the deformation constants are reasonably well known in some compounds,[142] the values of the Leslie coefficients have never been measured experimentally. In order to make the theory manageable, de Gennes assumes that all of the Leslie coefficients are of the same order of magnitude and have a value of the order of η, the viscosity. The Frank deformation constants which enter the eigenfrequency expressions can be taken to be comparable ($\sim K$) in magnitude for the nematic phase in which case the eigenfrequencies for the wave vector q and polarization $\alpha = (1,2)$ are written

$$\omega_{s\alpha} \equiv i\, U_{s\alpha} \sim i\, \frac{Kq^2}{\eta} \tag{18a}$$

$$\omega_{F\alpha} \equiv i\, U_{F\alpha} \sim i\, \frac{nq^2}{\rho} \tag{18b}$$

$$U_{s\alpha}/U_{F\alpha} \gg 1$$

where the subscripts s and F refer to the slow and fast modes and ρ refers to the density. As pointed out in the paper by the Orsay group, the expressions above may take on a different form if some of Leslie's coefficients turn out to be much larger than η.

The power spectrum defined as

$$I_\alpha(q,\omega) = (2\pi)^{-1} \int_{-\infty}^{\infty} dt\, e^{i\omega t} \langle \delta N_\alpha(-q,o)\delta N_\alpha(q,t)\rangle , \tag{19}$$

was calculated to be

$$I_\alpha(q,\omega) \cong \frac{kT}{\pi\gamma_1 \rho U_{F\alpha}} \left[\frac{P_\alpha}{\omega^2 + U_s} - \frac{C_\alpha Q_\alpha}{(\omega^2 + U_{F\alpha}^2)\gamma_1} \right]$$

(20)

where P_α and $C_\alpha Q_\alpha/\gamma_1$ are quantities comparable in magnitude ($\sim \eta q^2$) and γ_1 is the difference between two Leslie coefficients. If there is a fast and a slow eigenfrequency as in Equation 18, then the first term in the above expression is dominant and I_α is completely controlled by the slow mode in which case the integrated intensity becomes

$$\int I_\alpha(q,\omega) \, d\omega \cong kT/Kq^2 ,$$

(21)

which is the thermal amplitude of a mode of wave vector q. This result will be discussed in connection with spin-lattice relaxation.

The above results have been applied to light scattering by considering fluctuations in the dielectric tensor.[53,177] If the quantities ω_0, $\mathbf{k}_0$ and $\mathbf{i}$ are, respectively, the frequency, the wave vector and the polarization vector of the incoming beam and ω_1, $\mathbf{k}_1$ and $\mathbf{f}$ are these quantities for the outgoing beam, then the differential scattering cross section per unit solid angle, $d\Omega$, is written as

$$d\sigma/d\Omega d\omega = \pi\lambda^{-4}$$
$$\times \sum_{\alpha=1,2} I_\alpha(q,\omega)(i_\alpha f_o + i_o f_\alpha)^2 ,$$

(22)

where $f_o = \mathbf{f} \cdot \mathbf{N}_o$ and $i_o = \mathbf{i} \cdot \mathbf{N}_o$. The terms $i_\alpha = \mathbf{e}_\alpha \cdot \mathbf{i}$ and $f_\alpha = \mathbf{e}_\alpha \cdot \mathbf{f}$ where the $\mathbf{e}_\alpha$ are the orthogonal unit vector established by the wave vector $\mathbf{q}$ and $\mathbf{N}_o$ as diagrammed in Figure 24. In the above expression $\omega = \omega_0 - \omega_1$ and $\mathbf{q} = \mathbf{k}_0 - \mathbf{k}_1$.

In an order of magnitude calculation, de Gennes shows the cross section calculated above is dominant over a cross section determined from density fluctuations. In addition, it is further pointed out that from careful scattering experiments with the sample in and out of the magnetic field, it should be possible to obtain values for the deformation constants as well as for the Leslie coefficients.

It should be mentioned here that cholesteric liquid crystals show additional scattering phenomena which result from their helical structure. In this liquid crystalline phase a wavelength of maximum scattering occurs which is directly proportional to the pitch of the structure. That is, when exposed to white light the cholesteric structure scatters the light to give an iridescent color which varies with temperature and angle of incidence of the incoming light. These interesting systems have been studied by Fergason[121] and more recently by others.[123,179-182]

As mentioned earlier, the fluctuation modes described by de Gennes appear to play a strong role in spin-lattice relaxation. All of the work reported thus far has been on proton spins. In this case we are interested in determining T_1 as affected by the dipole-dipole interaction between nuclear spins modulated by the orientational modes. In order to make the calculation of T_1 manageable, it is necessary to idealize and consider the fluctuating interaction between only two proton spins. Even though there are many protons per molecule, this simplification of considering only the two nearest spins does not seriously affect T_1 and its frequency, temperature and orientation dependence. The coupling between two nuclear

FIGURE 24

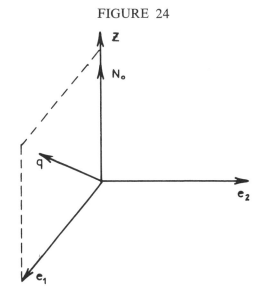

Definition of the unit vectors e_α which specify the fluctuation modes for the preferred direction of the nematic liquid crystal aligned along the z axis.

spins is described by the Hamiltonian

$$\mathcal{H} = \frac{(\gamma\hbar)^2}{r^2} \{(3n^2-1)[I_{1z}I_{2z} - \tfrac{1}{4}(I_1^+I_2^- + I_1^-I_2^+)]$$

$$+ \tfrac{3}{4}[(\ell+im)^2 I_1^+I_2^+ + CC]$$

$$+ \tfrac{3}{2}[(\ell+im)n(I_1^+I_{2z} + I_{1z}I_2^+) + CC], \quad (23)$$

where γ is the gyromagnetic ratio; r the internuclear separation; ℓ, m, n, the direction cosines of the internuclear vector with the z direction established by the direction of the magnetic field; I_z and $I^\pm$ being the appropriate spin operators. We are interested in fluctuations in ℓ, m, and n. How these quantities fluctuate depends upon how the molecule itself fluctuates. As described earlier, the long axes of the molecules are individually reorienting about some direction $N(r)$ in addition to collectively reorienting about the fixed direction N_0. An appropriate calculation of T_1 should include both motions. An exact calculation requires more information than we have at hand. Pincus,[183] however, showed that the frequency dependence of T_1 could be explained by considering only the collective motion and by attaching the direction $N(r)$ directly to the molecule. Using Equations 18a and 21 from the fluctuation theory T_1 becomes

$$T_1^{-1} = \omega_d \tfrac{3}{2} \frac{kT}{k} [\omega(D+K/\eta)]^{-\tfrac{1}{2}} \quad (24)$$

where ω_d is a constant depending on nuclear parameters and D is the diffusion constant. The diffusion term was introduced since D is of the order of K/η in many nematic liquids. The original expression, according to Pincus, contained the degree of order which has been set equal to one here. This expression predicts a $\omega^{\tfrac{1}{2}}$ dependence for T_1. Figure 25 shows an experimental plot of T_1 vs. ω by Doane and Visintainer[178] for p-azoxyanisole. The agreement with theory is reasonable; however, caution should be exercised since this result will not predict the temperature dependence. This might be expected in view of the assumptions used; however, a subsequent calculation of T_1 by Doane and Johnson[184] in whch the short-range fluctuations are considered in addition to the collective modes accounts reasonably well for the measured temperature dependence. The effect of short-range fluctuations introduces an S^2 term in the numerator of Equation 24. Since S^2/K, $(K/\eta)^{\tfrac{1}{2}}$ and $(D)^{\tfrac{1}{2}}$ have little dependence on temperature throughout the nematic range, T_1 has little temperature dependence. This is what is generally observed.[178] It is interesting to note that T_1 in smectic compounds likewise follow $\omega^{\tfrac{1}{2}}$ frequency dependence. It is disturbing, however, that no change is observed in T_1 in passing from the nematic to the smectic phase as Equation 24 would predict.

Blinc et al[185] repeated the calculation of Pincus, including a magnetic term in the fluctuations and neglecting the effect of diffusion. His final expression shows that $T_1 \propto \omega$ in the limit $\chi_a H^2/\eta \gg \omega$ where χ_a is the difference between the susceptibilities perpendicular and parallel to the magnetic field H. His expression, however, reduces to $T_1 \propto \omega^{\tfrac{1}{2}}$ in the limit $\chi_a H^2/\eta \ll \omega$. For p-azoxyanisole $\chi \cong 10^{-5}$, $\eta \cong 10^{-2}$ poise. A field of 10^4G gives $\chi_a H^2/\eta \cong 10^5 \text{sec}^{-1}$. An equivalent field gives a value of the Larmor precession frequency, ω, for protons of 10^8sec^{-1}. As originally suggested by Pincus, it would appear that the magnetic field would have little effect on T_1 which is, in fact, what Figure 25 suggests.

FIGURE 25

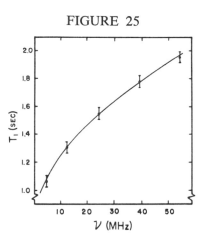

Frequency dependence of spin-lattice relaxation in p-azoxyanisole at 124°C. Solid line is the predicted $\nu^{\tfrac{1}{2}}$ dependence fitted to the measured values of T_1 at frequencies 12 and 24 MHz.[178] With permission.

It is clear that spin-lattice relaxation in liquid crystals is markedly different than that seen in normal liquids.[177,186] The agreement between theory and experiment for the frequency and the temperature dependence of T_1 indicates that order fluctuations are responsible for this difference.

IX. MAGNETIC RESONANCE

During the past decade a large fraction of published work on liquid crystals, more than 100 papers, has involved magnetic resonance studies. Perhaps the main reason for this interest is that magnetic resonance gives information on processes at the molecular level. These studies not only involve the liquid crystalline state but also molecules dissolved and ordered in liquid crystals.

The first magnetic resonance study of liquid crystals was reported in 1953.[187a,187b] Spence and his co-workers observed that the nmr spectrum of *p*-azoxyanisole in the nematic phase was considerably different than that seen in solids or in normal liquids. They correctly interpreted this spectrum in terms of order resulting from parallel orientation.[188,189] A detailed analysis of this spectrum came later when Lippman[190,191] and Weber[191-193] used a method of moments to show that the shape of the spectrum in the liquid crystal phase depended upon the degree of molecular order and the configuration of the molecule which comprised this state of matter. Magnetic resonance studies of solutes in nematic liquid crystals began with Saupe and Englert[194] who first recognized that it should be possible to observe the highly resolved spectrum of a simple solute molecule on top of the broad, unresolved spectrum from the liquid crystal solvent. They found that a wealth of information could be obtained about the ordered solute molecule including relative bond lengths and angles to a precision unequaled by other techniques, signs of spin-spin coupling constants and anisotropic chemical shifts. In addition, their research opened a new avenue in which to probe the liquid crystalline state. Since their pioneering work, a number of workers have become interested in this aspect of liquid crystals. Several excellent review papers on nmr of ordered solutes have appeared in the literature.[195-198] Recently, a number of research papers illustrating the properties of the liquid crystalline state have been published. Our review of the nmr of solutes will be primarily concerned with some of this recent work. For the reader who is interested in structural studies of solute molecules in nematic liquids, we refer him to the review papers indicated above and to the early papers of Saupe[199] and Snyder.[200]

Since the observations of Carrington and Luckhurst[201] there has been an increasing interest in electron spin resonance work in nematic liquids. Like high resolution nmr, this work has involved the use of solutes; however, because of the sensitivity of epr, much less solute is needed than in nmr. Recent epr work has shown the significance of this technique in the study of viscous liquid crystals.

Under the category of magnetic resonance one can also include the Mössbauer effect which has recently been observed in smectic liquid crystals.[232]

Nearly all the measurements made by the techniques mentioned above are sensitive to the mean orientation of the liquid crystal molecule or the solute molecule in the liquid crystal, whichever is being observed. A convenient way of describing this average orientation is through the use of the ordering matrix originally introduced by Saupe.[199] This matrix is defined in terms of a rectangular coordinate system with axes x_i ($i = 1,2,3$) fixed to the molecule and some preferred direction of orientation **N**. If ξ is the angle of each of the axes relative to **N**, then the matrix elements are given by

$$S_{ij} = \tfrac{1}{2}\langle 3\cos\xi_i \cos\xi_j - \delta_{ij}\rangle, \quad i,j = 1,2,3. \quad (25)$$

It is clear that this is a matrix of nine elements; however, it is symmetric as well as traceless and, when referred to principal axes, reduces to, at most, two elements. For an axially symmetric molecule the equation reduces to one element. The liquid crystalline molecules are typically long and if x_3 is chosen to be a long molecular axis, then the approximation can be made that $S_{11} \cong S_{22} \cong \tfrac{1}{2}S_{33} = \tfrac{1}{2}S$. In the case of high resolution nmr of solute molecules the above approximation cannot usually be made. However, if the molecule has a threefold axis of symmetry, the ordering matrix can be reduced identically to one element S.

The degree of order transforms as a second rank tensor which makes it possible to conveniently refer the order to any other rectangular coordinate system. The degree of order referred to an axis of the new (primed) system is then given by

$$S' = \sum_{ij} \cos\alpha_i \cos\alpha_j S_{ij}, \quad (26)$$

where the quantities $\cos\alpha_i$ ($i = 1,2,3$) are the direction cosines describing the axial direction in the molecular frame. If the molecule has axial symmetry, then Equation 26 reduces to

$$S' = \tfrac{1}{2}(3\cos^2\alpha_3 - 1)S. \quad (27)$$

Finally, it should be pointed out that a maximum value of $S = 1$ corresponds to ideal or perfect alignment and a value of $S = 0$ corresponds to a tumbling molecule. Values of S typically range from 0.3 to 0.7 for a liquid crystal molecule but may vary from zero to near 1.0 for a molecule dissolved in the liquid crystal.

A. Nuclear Magnetic Resonance

There are several interactions which may contribute to the nmr spectrum seen from a liquid crystal or from a compound dissolved and ordered in a liquid crystal. The interaction of primary interest is the magnetic dipole coupling between various nuclei within the molecule. For nuclei of spin ≥ 1 this may be dominated by the nuclear quadrupole interaction. In nematic liquid crystals of low viscosity, quadrupole effects appear to be observable only for nuclear spins such as the deuteron where the quadrupole moment is small. In magnetic fields with homogeneities of a few tenths of a gauss, typical of that used in wide-line resonance studies, the two interactions mentioned above are the only ones of significance in the shape of the nmr spectrum. In high resolution work, on the other hand, phenomena which result from interactions of the nuclei with electrons such as chemical shifts and indirect coupling of nuclear spins may become important.

The success of nmr in liquid crystalline systems is due to the rapid translational diffusion of the molecules in many of these phases. As a result, the nuclear magnetic dipole interactions between spins belonging to different molecules average to zero. The spectrum then depends only upon the interacting spins within the molecule. The fact that a broad nmr spectrum exists at all is a consequence of the partial molecular order exhibited by these compounds. The molecules on the average are parallel to some preferred direction. As a result of the anisotropy of the diamagnetic susceptibility, the preferred direction is established by the direction of an external magnetic field and made uniform over the whole sample.

Neglecting small nuclear-electron effects, the appropriate Hamiltonian describing a system of nuclear spins in the presence of an external field H_o written in the notation of Saupe is

$$\mathcal{H} = \sum_k (-\gamma_k \hbar H_o I_{zk})$$
$$+ \frac{\hbar}{2} \sum_{jk} B_{jk}(3I_{jz}I_{kz} - \underline{I}_j \cdot \underline{I}_k), \quad (28)$$

where γ_k is the gyromagnetic ratio of the kth nucleus, I the spin operator I_z the z component of the spin operator. The quantity

$$B_{ij} = -\frac{\hbar}{2\pi} \gamma_k \gamma_j \tfrac{1}{2} \langle (3\cos^2\theta - 1)/r_{jk}^3 \rangle. \quad (29)$$

where θ is the angle between an internuclear vector r and the direction of the magnetic field.

In a normal liquid, B_{ij} vanishes since an isotropic average of the term $\langle 3\cos^2\theta_{jk} - 1 \rangle$ is zero. In a nematic liquid crystal, B_{ij} can be expressed in terms of the degree of order by use of Equations 25 and 26. In principle, therefore, it should be possible to obtain a measure of the degree of order of the liquid crystal molecule from its nmr spectra, that is, if molecular bond lengths and angles are known from X-ray or electron diffraction data. A complication arises, however, in that molecules which show liquid crystallinity are typically large and contain a large number of proton spins. As a consequence, there exists a multiplicity of spin states which appears as a single broad line with some structure superimposed or under high resolution as a complex spectrum of lines. To obtain a measure of S for the liquid crystal molecule, it is, therefore, necessary to use a method of moments or to find a small liquid crystal molecule which shows a simple enough spectrum to be evaluated. With regard to the latter, an illustration of the measurement of S is given by Rowell, Phillips, Melby and Panar.[202] They selectively deuterated *p*-azoxyanisole to simplify the spectrum. In the nematic phase, *p*-azoxyanisole shows a spectrum of three unresolved lines. When the methyl groups were deuterated, they obtained two completely resolved lines (Figure 26) which they assigned to the interaction between the ortho protons in the phenyl ring. For two interacting spins, Equation 28 predicts a spectrum of two lines with a splitting in units of magnetic field of

$$\Delta H = \frac{3}{2} \frac{\gamma \hbar}{r^3} \langle 3\cos^2\theta - 1 \rangle . \qquad (30)$$

By approximating axial symmetry for the molecule, the time averaged term can be related to the degree of order by Equation 27 where α would be the angle between the internuclear vector and the major axis of the molecule. For the case of *p*-azoxyanisole, Phillips et al. assumed that an additional averaging occurs in that both halves of the molecule are reorienting about the *para*-axis in which case

$$\tfrac{1}{2}\langle 3\cos^2\theta - 1\rangle = (\tfrac{3}{2}\cos^2\gamma - \tfrac{1}{2})(\tfrac{3}{2}\cos^2\Phi - \tfrac{1}{2})S, \qquad (31)$$

where γ is the angle between the internuclear vector and the *para*-axis, not to be confused with the gyromagnetic ratio, and Φ is the angle between the *para* and molecular axis as indicated in Figure 26(a). A value of $r = 2.45\text{Å}$ gives a value of 5.76G for $3\gamma\hbar/r^3$. For $\Phi = 10°$ and $\gamma = 0$ for the ortho protons, the splitting in Figure 27 gives a value of $S = 0.55$. The small splitting of 0.27G can be assigned to the interactions between the 2,6 and 3,5 aromatic protons.

FIGURE 26a

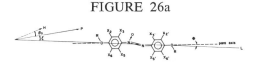

Reorientation axes, symbols, and angles for *p*-azoxyanisole.

FIGURE 26b

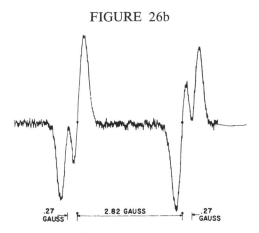

Corresponding proton resonance spectrum at 120°C in the nematic phase at 60 MHz in which the methyl groups, R, have been deuterated. (Modified from Reference 202). With permission.

FIGURE 27

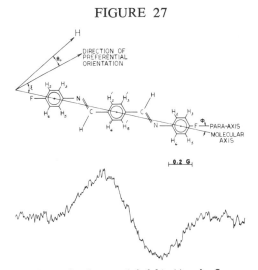

The molecule terephthal-*bis*-(4-aminofluorobenzene) and fluorine resonance in the nematic phase at 163°C. (Modified from Reference 203). With permission.

For cases where there is no resolved structure on a wide line spectrum, the method of moments has been used. The second moment is given by[191,203]

$$\langle \Delta H^2 \rangle = \frac{\int_0^\infty (H-\langle H \rangle)^2 f(H) dH}{\int_0^\infty f(H) dH}, \quad (32)$$

where $f(H)$ describes the shape of the resonance line and H is the magnetic field at which resonance occurs. This quantity can be obtained numerically from the resonance line. Theoretically, it can be found from the expression developed by Van Vleck

$$\langle \Delta H^2 \rangle = \frac{3}{4} \frac{I(I+1)}{\gamma^2} \frac{4\pi^2}{N} \sum_{pq} B_{pq}^2$$

$$+ \frac{1}{3} \frac{I_F(I_F+1)}{\gamma^2} \frac{4\pi}{N} \sum_{pF} B_{pF}^2 \quad (33)$$

where the summation in the first term is over spins of the same species as the observed spin and the summation in the second term over unlike spins with a spin I_F.

An example[203] of the use of this quantity is given for the case of a Schiff's base in Figure 27. The protons do not provide a good means for the determination of the degree of order in this case because of uncertainties in the configuration of the molecule. The shape of the proton spectrum depends strongly on the position or motion of the proton in the amine linkage relative to the aromatic rings. Confusion exists in the literature concerning this positioning and, thus, the effective bond lengths are uncertain. The fluorine resonance, on the other hand, is independent of the relative position of the amine proton and is far enough away from the central ring to be little influenced by its proton spins. The fluorine resonance could then be used to obtain a value of S. The curve in Figure 27 gives a value of $(\Delta H^2) = 2.70 \times 10^{-2}$ G². Bond distances and angles can be found in the literature and the use of Equations 29, 31 and 33 gives a value of $S = 0.58$. With known values of S the second moment of the proton line was used to study the molecular configuration.[203]

Values of degree of order measured by these methods are in general agreement with values determined by other techniques as well as with their theoretically predicted values from the statistical theory of Maier and Saupe.[6] This theory, based upon dispersion interactions is discussed earlier in the section on thermodynamics. The theory agrees with the measured temperature dependence of S for low values of the reduced temperatures, but near the isotropic transition the fit is generally not as good. In addition to dispersion interactions, a number of other possible factors have been considered which could influence the molecular order such as molecular shape, permanent electric dipole moments, etc. In order to determine the relative contributions of each of these effects the use of solutes in liquid crystals has been employed.[204-207] The use of solutes not only offers more control over the experimental conditions but also the order parameters of simple solutes can be measured with extreme precision with high resolution techniques. Values of S are typically quoted to the third and fourth place after the decimal.

In order to determine the relative significance of permanent electrical dipole moments and dispersion type forces, Saupe[204] has performed some experiments on the chlorobenzenes and, later, Nehring and Saupe performed experiments on fluorobenzenes.[205] To study the effect of the electric dipole, Saupe dissolved the compound, 1,2,3-trichlorobenzene, which possesses a permanent electric dipole moment, in a number of solvents with positive and negative dielectric anisotropies. It was expected that there would be a large variation in the relative average orientation since the electric dipole moment will tend to be in the direction of the largest dielectric susceptibility. Little variation was observed, indicating dipole-dipole forces are of minor importance.

Nehring and Saupe[205] have studied the orientation of several selectively fluorinated benzenes in a common solvent. They observed differences in the orientation of different fluorobenzenes. Regarding dispersion type forces, these differences could be attributed to localized contributions of the fluorine substituted bonds. Similar conclusions were also obtained for chlorobenzenes even though the larger chlorine atom has more influence on the shape of the molecule. The effect of molecular shape and concentration will be discussed later in connection with epr studies.

When a nematic liquid crystal is placed in a magnetic field it tends to align so that its axis of minimum susceptibility is parallel to the external field. Recently, several workers have been interested in orientational studies, that is, finding methods to alter the direction of preferential alignment relative to the magnetic field direction. This has been accomplished in several ways: use of smectic materials, applied a.c. and d.c. fields and use of viscous materials.

One of the most interesting methods for orientation is the use of smectic materials. Yannoni[208] has succeeded in ordering a solute molecule in a smectic A texture. Figure 28 shows the fluorine spectra for a 20 mole % solution of 1,1,1-trifluorotrichloroethane in the smectic phase of 4-(2-n-propoxyethoxy-benzylideneamino)acetophenone. An interpretation can be made based on Equation 29. Since the solute molecule can be assumed to be a cylindrically symmetric environment about its preferred axis, Equation 29 can be altered thus—

$$B_{ij} = -\frac{\hbar}{2\pi} \gamma_k \gamma_j (\frac{3}{2} \cos^2 \theta_o - \frac{1}{2})$$

$$\times \frac{1}{2} \langle (3\cos^2\theta'_{jk} - 1)/\gamma_{jk} \rangle , \quad (34)$$

where θ_o is the angle between the preferred direction of alignment and the magnetic field direction and θ' is the angle between the preferred axis and the internuclear vector. The smectic A texture was ordered by cooling the sample from the isotropic melt in the presence of a magnetic field. Once ordered, the sample would retain the order and could be rotated (i.e., θ_o varied) without further being influenced by the field. The splittings in Figure 28 are seen to vary as $(\frac{3}{2}\cos^2\theta_o - \frac{1}{2})$ which is predicted by Equation 34. Even removing the sample from the field for 96 hours did not affect its order; the temperature must remain constant.

From the splittings, the degree of order for the smectic phase was determined. The Hamiltonian (Equation 28) has been solved for a three spin system. Use of Equation 34 then gives the splitting for this case as

$$\Delta\nu = \frac{3}{2} \frac{\gamma^2 \hbar}{2\pi r^3} (\frac{3}{2}\cos^2\theta_o - \frac{1}{2})S. \quad (35)$$

Yannoni measured a value of S = 0.0148 which is similar to the value for this solute in nematic liquids.

Carr and his co-workers[102,209] have extensively studied the effect of a.c. electric fields

FIGURE 28

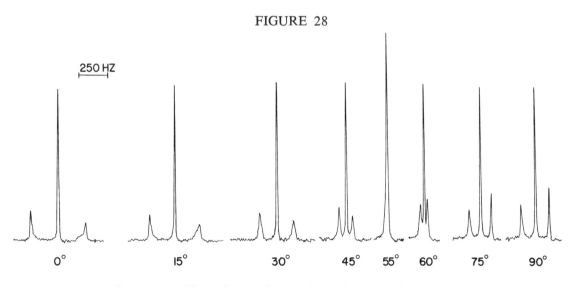

NMR spectra of ^{19}F in 1,1,1-trifluorotrichloroethane oriented in a smectic A liquid crystal solvent. The spectra were obtained at various angles with respect to the initial direction (0°) along which the sample oriented.[208] With permission.

on nematic liquid crystals. Figure 29 shows their proton spectra for 4-(anisalamino)phenyl acetate with no electric field applied (b) and with a 300-kHz field of 1500 V/cm applied parallel to the magnetic field (c). The splittings could be assigned to groups making up the molecule. Carr found that with the application of a 300-kHz electric field parallel to the magnetic field direction, the splittings between each pair of satellites reduced by a factor of ½ (i.e., 3.3G to 1.6G and 6.7G to 3.4G). Equation 34 indicates a 90° rotation of the preferred direction of alignment. This rotation was expected since the largest dielectric constant is perpendicular to the magnetic field. They further found, however, that upon lowering the frequency of the applied electric field into the audio range, the preferred direction of alignment was altered from perpendicular to parallel to the direction of the applied field. A loss in resolution was also observed at the low frequencies. This behavior is not altogether understood. Based upon their measurements of dielectric loss at microwave frequencies, Carr and Twitchell[99] have suggested a process which involves polarization effects due to conductivity in the sample. They also point out a dependence on impurities.

When the electric and magnetic fields are applied simultaneously, they have a competitive influence on the orientation of the molecule along a preferred axis. At the appropriate field strengths the orientation becomes random. Values of (E/H) ratios for random orientation at low frequencies were measured by Carr and Twitchell to be near 0.4–0.6 V/cm G. For further discussion of these studies see the section on external forces and fields.

Recently, Diehl et al.[210] reported uniform alignment of the nematic phase with d.c. electric fields. They observed the high resolution spectra of cis-1,2-dichloroethylene dissolved in n- propyl 4-methoxybenzylidene-4-amino-α-methylcinnamate at 27°C and aligned with an external d.c. electric field shown in Figure 30. In their sample they report the specific re-

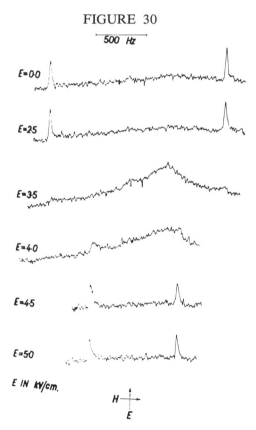

FIGURE 30

Influence of electric fields on the NMR spectrum of cis-1,2-dichloroethylene dissolved in a room temperature nematic. Concentration = 22 mole percent; temperature 27°C.[210] With permission.

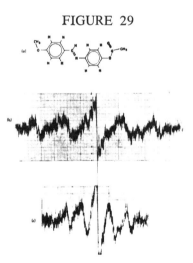

FIGURE 29

PMR spectra of 4-(anisalamino) phenyl acetate at 90°C and 6.4 MHz. (a) Structural formula. (b) NMR spectrum with no applied electric field (E = 0; H = 1500 G). (c) NMR spectrum with a 300-kHz electric field of 1500 V/cm applied parallel to the external magnetic field (E = 1500 V/cm, H = 1500 G, and E ∥ H).[209] With permission.

sistance to be high $\sim 10^9 - 10^{10}$ Ω/cm, indicating small conduction effects.

Viscous materials have also been reported to have been used to vary the direction of the preferred axis relative to the magnetic field.[211] Figure 31 shows the proton spectrum of methylene chloride in poly-γ-benzyl-L-glutamate for various residence times in the magnetic field. The bottom spectrum is a typical powder pattern expected for random alignment of the preferred axis. After several hours in the field, the molecules become oriented. Once aligned, the sample can be rotated.

Sackmann, Meiboom and Snyder[127][129][197][212] have performed some interesting experiments with the cholesteric phase. They have shown that it is possible to observe the spectrum of a solute molecule when dissolved in a cholesteric liquid crystal that orients with its pitch axis parallel to the external magnetic field. This appears to be particularly true for compensated mixtures of cholesteryl derivatives of opposing helicities. They were also able to show that in the case of compensated mixtures, that is, no macroscopic helicity, a solute molecule may still experience a local helical environment since the solute spectra in the mixed cholesteric solvents differs from that seen in nematic solvents.

When the pitch axis is aligned perpendicular to the field, they were able to induce a cholesteric-nematic transition with the applied magnetic field. Figure 32 shows the spectrum of benzene dissolved in a mixture of optically active n-amyloxyazoxybenzene and 4,4'-di-n-heptyloxyazoxybenzene. The spectrum vanishes at 100°C (cholesteric phase) but is present at 106°C (nematic phase). The process is reversible.

Recently, Carr, Parker and McLemore[213] have studied the effects of electric fields on mixtures of nematic and cholesteric liquid crystals. With the application of an a.c. electric field they observe a cholesteric-nematic transition.

Lyotropic liquid crystal phases have been studied by nmr.[214] One of the most interesting of these phases has been reported by Flautt, Black and Lawson.[215][216] This particular phase is formed by a mixture of C_8 or C_{10} alkyl sulfates, the corresponding alcohol, sodium sulfate and water in proportions of 40, 5, 5, and 50, respectively. The phase will orient in a magnetic field and show a temperature range from 10–75°C. The material is viscous and will orient slowly when placed in a magnetic

FIGURE 31

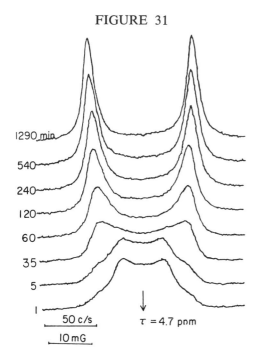

Variation of the doublet of methylene chloride in liquid crystalline state with the residence time of the sample in the magnetic field.[211]

FIGURE 32

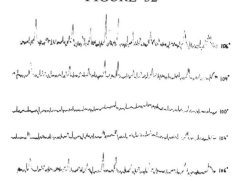

NMR spectra at 60 MHz of a mixture of 0.013 g of optically active amyloxyazoxybenzene, 0.574 g of hexyloxyazoxybenzene, and 0.03 g of benzene. The spectra were obtained in succession, from top to bottom, at the temperature indicated at right. The small sharp peaks, most distinct in the center spectrum, are frequency markers spaced 106 Hz apart. Total width of the spectra is about 2000 Hz.[127] With permission.

field. This is evidenced by the deuterium resonance (see Figure 33); the powder pattern indicates random order for the sample when initially placed in the field and the two-line spectra are expected for a spin one quadrupole interaction such as deuterium in D_2O where the molecules are uniformly aligned throughout the sample.

The most interesting aspect of this phase, however, is that it will order solute molecules perpendicular to the magnetic field. This results in an unusual effect when the sample is first rotated in the magnetic field. If after the sample is placed in the field and allowed to order, then rotated 90°, a powder pattern will result. If the sample is allowed to reorder and is again rotated 90°, the powder pattern will be diminished. Succeeding rotations will eventually completely order the solute. Black, Lawson and Flautt have explained the 90° alignment of a solute in terms of the structure of the surfactant aggregates.

Flautt and Lawson[217] have studied the more common lyotropic phases, neat, middle and viscous isotropic. Among other lyotropic systems they have studied the proton and deuteron spectra in dimethyldodecylamine oxide $[CH_3(CH_2)_{11}N(O)(CH_3)_2]$-deuterium oxide $[D_2O]$ system ($DC_{12}AO$-D_2O). In the neat and middle phases the deuterium spectra of D_2O show a typical powder pattern. The powder pattern reduces to a narrow line in the fluid isotropic and viscous isotropic phases. The proton spectra show a broad Lorentzian shaped line. This unusual shape has been attributed to a distribution in correlation times. They have been able to distinguish the neat and middle phases from these lines.

In Figure 33, one might be tempted to determine the degree of order from the splitting of the deuterium spins. The quadrupole Hamiltonian for a spin one is formally equivalent to the dipole-dipole interaction Hamiltonian Equation 28 for two interacting spins provided that $\gamma^2/\hbar^2/r^3$ is replaced by $\frac{1}{2}e^2/qQ$ where q is the electric field gradient and Q is the quadrupole moment of the nucleus. The angle θ must now describe the direction of the field gradient. Equation 30 now becomes

$$\Delta H = \frac{3}{4} \frac{e^2qQ}{\gamma\hbar} <3\cos^2\theta - 1> , \qquad (36)$$

for the splitting between spectral lines. An axially symmetric field gradient has been assumed. A similar problem arises, as in the case of proton resonance, in that one is not always sure how much averaging occurs as a result of molecular rotations. Also, the magnitude of the quadrupole coupling constant may be uncertain. Equation 36, however, has been used to measure order in thermotropic systems[202] and of ordered solutes.[206]

B. Electron Paramagnetic Resonance

A technique which has been gaining increasing interest in liquid crystal studies is that of electron paramagnetic resonance (epr). The compounds which show liquid crystallinity are not, in general, paramagnetic. The technique is, therefore, limited to solute studies where the solute molecule contains an unpaired electron spin. The concentration of the solute required for epr is considerably less than that needed for nmr work. Mole fractions less than 10^{-3} give observable spectra, the ultimate reason for the increased sensitivity being the large value for the Bohr magneton (10^3 that of the nuclear magneton).

This sensitivity can be particularly advantageous in probing liquid crystals since many of the properties of this state of matter are strongly dependent on dissolved foreign substances. Electron paramagnetic resonance (EPR) studies of the nematic phase typically show that required concentrations of the para-

FIGURE 33

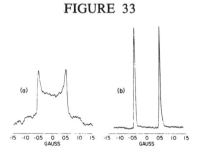

Deuterium spectra of a system containing 50% D_2O, 36% SDS, 7% DeOH, and 7% Na_2SO_4: (a) Immediately after being placed in the magnet; (b) after being in the magnet for about 45 minutes.[215] With permission.

magnetic probe do not significantly alter such properties as the nematic-isotropic phase transition temperature or the width of the transition.

Another advantage of epr is that the Hamiltonian which describes the observed spectrum contains interactions which are intramolecular. That is, effects of the neighboring solvent molecules do not enter the expression even in viscous materials where the absence of motional narrowing makes nmr studies prohibitive because of intermolecular interactions. This property will be discussed later in connection with recent measurements of order in viscous liquid crystals. The effect of the liquid crystal "lattice" on epr at all temperatures is the same as that of nmr at high temperatures, namely that of partial ordering and thermal modulating the probing molecule.

The first reported work of epr in liquid crystals was that of Carrington and Luckhurst,[201] who observed diphenylpicrylhydrazyl (DPPH) and the tetracyanoethylene anion (TCNE⁻) dissolved in p-azoxyanisole. At temperatures above the nematic-isotropic transition they observed five- and nine-line spectra, respectively, which were characteristic of that seen in any nonviscous isotropic liquid crystal. Upon lowering the temperature into the nematic phase, the splittings between the lines were observed to decrease abruptly at the transition, then continue to decrease continuously as the temperature was further lowered through the nematic phase. This observation was predicted by Carrington and Luckhurst and correctly interpreted in terms of the partial alignment of the solute molecule induced by the host nematic.

To review the epr work in liquid crystals it is necessary, first of all, to illustrate the technique. We take as an example vanadyl acetylacetonate (VAAC). This molecule contains the VO^{+2} radical and has been the most popular probe in recent investigations.[218-225] The appropriate axial spin Hamiltonian referred to the laboratory frame can be written as:

$$\mathcal{H} = g\beta H_z S_z + a\mathbf{I}\cdot\mathbf{S}$$
$$+ \tfrac{1}{3}(\Delta g\beta H_z + bI_z)(3\cos^2\theta - 1)S_z$$
$$+ \tfrac{b}{2}\sin\theta\cos\theta\,(I_+ e^{-i\phi} + I_- e^{+i\phi})S_z$$
$$+ \tfrac{1}{2}(\Delta g\beta H_z + bI_z)\sin\theta\cos\theta\,(S_+ e^{-i\phi} + S_- e^{+i\phi})$$
$$+ \tfrac{b}{4}\sin^2\theta(I_+ S_+ e^{-2i\phi} + I_- S_- e^{2i\phi})$$
$$- \tfrac{b}{12}(3\cos^2\theta - 1)(I_+ S_- + I_- S_+), \quad (37)$$

where the z direction of the laboratory frame is established by the direction of the magnetic field; the angles θ and ϕ being the polar and azimuthal angles referring to the V=O bond direction from z direction; $b = A_\parallel - A_\perp$, the difference between the parallel and perpendicular components of the hyperfine tensor; $a = \tfrac{1}{3}(A_\parallel + 2A_\perp)$; $\Delta g = g_\parallel - g_\perp$ the difference between the parallel and perpendicular component of the "g" tensor; $g = \tfrac{1}{3}(g_\parallel + 2g_\perp)$; S_z and I_z being z components of the electronic and nuclear spin operator; and $I_\pm$ and $S_\pm$ are the raising and lowering operators.

To apply this Hamiltonian to a VAAC molecule in a nematic liquid crystal, it is necessary to average over the angles θ and ϕ. The manner in which this average is performed depends upon the rate and manner in which the molecule is thermally agitated. The probe molecule is partially ordered about some direction determined by the host nematic. We may view the long axis of the molecule as vibrating in a random manner about this direction with a correlation time τ_c. τ_c is roughly a measure of the time over which there persists some correlation within an assembly of molecules. Depending upon the magnitude of τ_c, the average over θ and ϕ is either a spatial or a time average. As a consequence, two possible epr spectra result corresponding to the two extreme values of τ_c; $\tau_c < 1/\Delta\omega$ and $\tau_c \gg 1/\Delta\omega$

where $1/\Delta\omega \sim 10^{-9}$ sec, $\Delta\omega$

being the splitting between the hyperfine lines in units of angular frequency of the epr spectra. Fryberg and Gelerinter[218] have studied VAAC in a long temperature range nematic where both conditions can be made to exist. The compound they studied was N,N-(di-n-octyloxybenzylidene)-2-chloro-1,4-phenylenediamine which exhibits the nematic phase over the temperature range from 179°C to a supercooled temperature near 40°C. Their spectra are shown in Figure 34. The spectra above the nematic range are typical of that seen in an isotropic liquid. The presence of eight lines (hyperfine structure) is in accordance with the nuclear spin of V^{51}, $I = 7/2$, and the first two terms in Equation 37. In the isotropic case, the angular dependent terms average to zero. In the nematic phase above 105°C the condition $\tau_c < 1/\Delta\omega$ is obeyed. In this case the Hamiltonian in the high field approximation which describes the spectra becomes

$$\mathcal{H} = g\beta H_z S_z + a\mathbf{I}\cdot\mathbf{S}$$
$$+ \frac{1}{3} bI_z \langle 3\cos^2\theta - 1 \rangle S_z, \quad (38)$$

where Δg has been taken to be zero and the brackets on the term $\langle 3\cos^2\theta - 1 \rangle$ represent a time average. The off-diagonal terms in Equation 37 do not contribute.

Upon finding the eigenstates, Equation 38 predicts a spectrum with a variable splitting between the hyperfine lines and with an overall spectrum width, ΔH, less than that seen in the isotropic liquid. In order to eliminate second order terms which are responsible for the variation in the hyperfine splitting, it is common procedure to define an average splitting $\langle a \rangle = \Delta H/7$ for the case of spin 7/2 V^{51}. In this case Equation 38 predicts

$$\langle a \rangle = a + \frac{b}{3}(3\cos^2\theta - 1). \quad (39)$$

Using the definitions of a, b and the degree of order S, Equation 39 becomes

$$S = \frac{1}{2}\frac{\langle a \rangle - a}{a - A_\perp} \quad (40)$$

The values of a and $A_\perp$ (−107.3 and −68.5 G, respectively, for VAAC) are found from the spectra in the isotropic liquid and, hence, S can be determined. In the case of VAAC the V=O bond is perpendicular to the long axis and, hence, the preferred direction of orientation in which case S will be less than 0.5 and have a negative value.

Fryberg and Gelerinter found that the temperature dependence of S determined in this manner followed that expected by the Maier-Saupe theory down to 105°C (Figure 35) where S began to increase sharply and the appearance of the spectrum began to approach that of a "glassy" type. A similar observation was made by Diehl and Schwerdtfeger[219] in a room temperature nematic. They showed that this could be explained if the thermal motion of the molecules were approaching the condi-

FIGURE 34

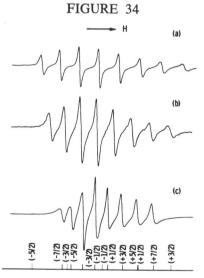

Experimental derivative spectra of vanadyl acetylacetonate in N,N-(di-n-octyloxybenzylidene)-2-chloro-1,4-phenylenediamine taken at: (a) A temperature (202°C) above the clearing point showing a spectrum characteristic of an isotropic liquid; (b) a temperature (155°C) just below the clearing point showing a spectrum characteristic of moderate order; (c) a temperature (51°C) just above the solidification point showing a spectrum characteristic of apparent high order.[218] With permission.

FIGURE 35

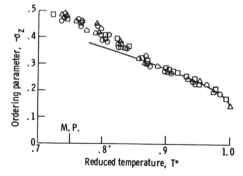

Plot of the ordering parameter, $-\sigma_z$, versus the reduced temperature, T°.[218] With permission.

tion $\tau_c \gg 1/\Delta\omega$. In this case it is necessary to perform a spatial rather than a temporal average. The calculation of a spatial average, however, demands a knowledge of the angular distribution of the molecules. Without actually carrying out this calculation, Gelerinter showed that the spectrum could be explained in terms of what one would obtain if the angular distribution were isotropic. In a viscous isotropic material where the long axis of the molecules are randomly ordered, epr spectra will show two sets of lines. One set of the lines in the spectrum will have a splitting nearly identical to that predicted for the bond direction perpendicular to the magnetic field direction, $H_\perp$, (solid lines in Figure 34) and the other set with a splitting parallel to the field direction, $H_\parallel$, (dotted line of Figure 34). As would be expected in a liquid crystal, there are no observed lines (Figure 34) for $\theta = 0$ (VAAC molecular axis at 90°) whereas there are spectral lines which correspond to $\theta = 90°$ as indicated by the solid lines in Figure 34.

Diehl and Schwerdtfeger[219,220] developed this idea further to perform a very clever experiment. They reasoned that by observing the intensity of the hyperfine lines at $H_\parallel$ and by rotating the preferred molecular direction by use of an electric field, it should be possible to determine the orientation distribution function. In their experiment they used VAAC dissolved in *n*-propyl-4'-methoxybenzylidene-4-amino-α-methylcinnamate which shows the nematic phase at room temperature. They found that by applying a transverse electric field of 2 kv/cm, the preferred direction of the bond axis could be made parallel to the magnetic field direction as indicated by the spectrum in Figure 36. By varying the angle between the electric and magnetic field, θ', the epr spectrum as a function of θ' ($\theta' = 90° - \theta$) could be determined. The electric field was rotated about an axis parallel to the rotating microwave field $H_\perp$ so that the induced transition probability was not affected by the rotation.

With the use of the solution to the Hamiltonian for the "glassy" state, Diehl was able to arrive at a simple relationship for the intensity of the hyperfine lines I at the field $H_\parallel$ in terms of the orientation distribution function $f(\theta)$. Diehl found that

$$I = \left(\frac{dN}{dH}\right)_{orient} = \left(\frac{dN}{dH}\right)_{random} f(\theta), \quad (41)$$

where dN is the number of VAAC molecules which have their bond axis oriented between the angles θ and $d\theta$. The experimental results for the signal intensity as a function of θ' normalized to one for $\theta' = 90°$ are shown in Figure 37. Diehl and co-workers compared their re-

FIGURE 36

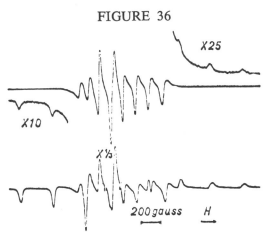

(a) ESR spectrum of vanadyl acetylacetonate in a liquid crystal at room temperature. The inserts show the wings of the spectrum under high amplification. (b) The same spectrum upon application of a transverse electric field of 2kv/cm.[219] With permission.

FIGURE 37

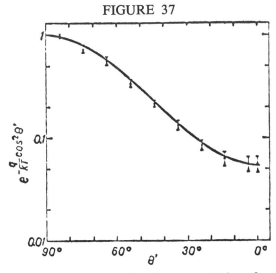

A plot of the intensity of the parallel hyperfine lines as a function of the angle θ' between the optic axis of the liquid crystal and the static magnetic field direction for vanadyl acetylacetonate dissolved in a liquid crystal at room temperature. The intensities are normalized to 1 at 90°. The solid curve is calculated for q/kT = 2.9.[220] With permission.

sults to an expression developed by Saupe for $f(\theta)$, i.e., $f(\theta) = \exp[-(q/kT)\sin^2\theta]$. The solid line in Figure 37 shows a plot of the function $I_{\shortparallel}(90°)\exp[-(q/kT)\cos^2\theta']$ with $q/kT = 2.9 \pm 0.1$. The agreement is remarkable. To the authors' knowledge, this is the first direct measurement of the distribution function.

Diehl and Schwerdfeger further showed how the ordering parameter in viscous liquid crystals could be determined with the use of the Saupe distribution function. In this case the value of q/kT is determined as above or, more simply, by the signal intensity ratio at $\theta = 0$ and $\pi/2$. By using the Saupe expression the average $S = \frac{1}{2} <3\cos^2\theta - 1>$ is calculated in terms of q/kT. With a value of $q = 0.07$ ev, Diehl obtained a value of $S_\perp = -0.27$ at room temperature.

Another interesting study involving the orientation distribution function is the paramagnetic relaxation work of Glarum and Marshall.[224,226] They calculated the transverse relaxation time, T_2, for the case of a nematic liquid crystal and compared their results with the spectra of several vanadyl complexes in *p*-azoxyanisole.

Experimentally, the quantity T_2 is proportional to the reciprocal of the line width. Physically, it is dependent upon fluctuations of the spin Hamiltonian. In order to calculate T_2, Glarum and Marshall found it necessary to average over the Legendre functions in the off-diagonal terms in the spin Hamiltonian. These terms involve $<\cos^2\theta>$ and $<\cos^4\theta>$. The former average can be measured; that is, it can be expressed in terms of the degree of order S which is determined from the hyperfine splittings. The average $<\cos^4\theta>$, on the other hand, must be calculated. In order to compute this average they used the same orientation distribution function $e^{\alpha\cos^2\theta}$ suggested by Saupe and as expressed earlier. The final result of the calculation of the line width, $1/T_2$, could be expressed in terms of a second order polynomial in the quantum number m, i.e.,

$$1/T_2 = A + Bm + Cm^2. \qquad (42)$$

The coefficients B and C are plotted in Figure 38 as a function of the degree of order ex-

FIGURE 38

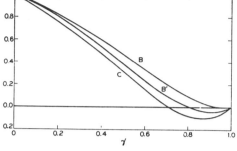

Variation of the B and C coefficients in Equation 42 with the degree of order. The curve B' includes second-order contributions to B.[224] With permission.

pressed as γ in the Glarum and Marshall paper.

An interesting feature of these curves is that the quadratic coefficient vanishes at $\gamma = 0.71$. It is negative for high degrees of order $\gamma > 0.71$ and positive for values of $\gamma < 0.71$. Since the spectral lines which occur at a low magnetic field correspond to negative values of m (m = $-\frac{7}{2}, -\frac{5}{2} \ldots +\frac{7}{2}$), Equation 42 predicts the following features of the spectra.

(a) For low degrees of order ($\gamma < 0.71$) the outer lines (large m values) will be the broadest with the high-field lines being broader than the low-field lines.

(b) When $\gamma \simeq 0.71$ the quadratic term vanishes and the lines become progressively broader in a linear fashion as the resonance fields increase.

(c) For $\gamma > 0.71$ the outermost lines become narrow relative to the center of the spectrum.

The agreement with experiment is phenomenal. Figure 39 shows all three cases for vanadyl dibenzoylmethane in *p*-azoxyanisole with spectrum a being in the isotropic liquid $\gamma = 0$, spectrum b for $\gamma \approx 0.7$ and spectrum c for $\gamma > 0.7$. Another example of case (a) can be seen in Figure 34. Glarum and Marshall found these results to be valid for VO^{2+} complexes with the β diketones listed in Table 8. *p*-Azoxyanisole and 4,4'-di-*n*-hexyloxyazoxybenzene were the solvents.

Results for the degree of order for these solutes are shown in Figure 40. The molecular order is clearly dependent on the molecular

FIGURE 39

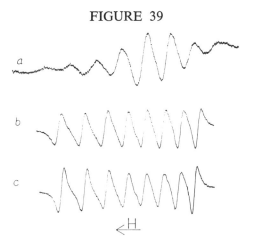

ESR spectra of the vanadyl dibenzoylmethane complex in *p*-azoxyanisole at 150°C. (a) 150°C; (b) 130°C; (c) 113°C.[224] With permission.

TABLE 8

Structures of the β-diketones, R_1—CO—CHR$_3$—CO—R$_2$

Chelate	R_1	R_2	R_3
2,4-Pentanedione	CH$_3$	CH$_3$	H
Benzoylacetone	C$_6$H$_5$	CH$_3$	H
Dibenzoylmethane	C$_6$H$_5$	C$_6$H$_5$	H
Di-2-naphthylmethane	C$_{10}$H$_7$	C$_{10}$H$_7$	H
3-Phenyl-2,4-pentanedione	CH$_3$	CH$_3$	C$_6$H$_5$
3-*n*-Octyl-2,4-pentanedione	CH$_3$	CH$_3$	C$_8$H$_{17}$

FIGURE 40

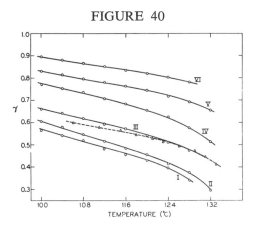

Variation of the degree of order with temperature of vanadyl complexes with: (I) 3-phenyl-2,4-pentanedione; (II) 3-*n*-octyl-2,4-pentanedione; (III) 2,4-pentanedione; (IV) benzoylacetone; (V) dibenzoylmethane; (VI) di-2-naphthoylmethane. The degree of order of *p*-azoxyanisole is given by the broken curve.[224] With permission.

shape and size. For large planar molecules it is seen that the order of the solute is greater than the solvent (dashed lines). Large molecules which are twisted from the planar structure are not as well aligned.

Chen, James and Luckhurst[223] have analyzed these data based on the statistical theory of Maier and Saupe and find that their calculations qualitatively account for the deviations in order where these different solutes are aligned in the same nematic solvent. Luckhurst and Chen[221] have studied the effect of solute concentration on the order of the nematic phase. They measured the order with a paramagnetic solute of low concentration, then introduced other solutes with varying concentration. Their results are seen in Figure 41 plotted against reduced temperature. It is seen that the presence of the solute does not seriously disturb the order of the nematic phase. The epr experiments mentioned so far have all used a vanadyl complex for the unpaired spin. Other paramagnetic probes[227-231] which have been used include the tetracyanoethylene anion,[201] Coppinger's radical, perinaphthenyl, *bis*(2,2-6,6-tetramethyliperidyl oxide-4-yl) glutamate

FIGURE 41

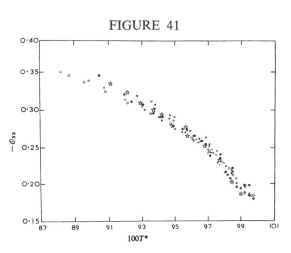

The degree of order $-\phi_{33}$ as a function of the reduced temperature T* for *p*-azoxyanisole and its mixtures. O, *p*-azoxyanisole: ▼, + ethyl alcohol 10.9 mole %; □, + benzene 5.13 mole %; ☆ + benzene 10.0 mole %; ▽, + benzene 16.0 mole %; ●, + CCl$_4$ 3.9 mole %; ◇, + CCl$_4$ 7.1 mole %; ■, + phenanthrene 3.85 mole %; △, + phenanthrene 4.61 mole %; ▲, + phenanthrene 7.17 mole %; ▷, + *o*-terphenyl 5.56 mole %; △, + *o*-terphenyl 7.61 mole %; ◀, + *o*-terphenyl 5.56 mole %; ◆, + phenanthrene 4.04 mole %.[221] With permission.

and *bis*(2,2,6,6-tetramethylpiperidyl oxide-4-yl)terephthalate.

Luckhurst[231] has been interested in using epr to distinguish between theories of the nematic liquid crystal state, i.e. between the swarm and continuum theory. He has written the spin Hamiltonian for the case of the swarm model. In this case he visualizes large groups of molecules grouped into swarms with their long axes parallel. The swarms were then surrounded by molecules moving about in an isotropic fashion. Within the swarms there are an x number of solute molecules. An ordering matrix S_{ab}^A describes the order of these molecules since the clusters themselves were only partially ordered. Outside the clusters there were 1-x solute molecules with the rate of exchange between the two regions being rapid enough to write a time averaged Hamiltonian. Luckhurst found that the Hamiltonian that resulted contained the quantity xS_{ab}^A and could be made identical to the usual Hamiltonian by exchanging the ordered matrix S_{ab} with xS_{ab}^A; hence, the two models are undistinguishable.

C. Mössbauer Effects

Of the most recent investigations in liquid crystals, one of the most spectacular and significant is that of the observation of the Mössbauer effect in these phases. Uhrich et al.[232] observed the quadrupole split ^{57}Fe spectrum of 1,1'-diacetylferrocene dissolved in 4,4'-di-*n*-heptyloxyazoxybenzene in the smectic C texture. This is the only Mössbauer work in liquid crystals that has been reported to date. A brief description of this work and its implications will be given.

The Mössbauer effect is based upon recoil-free emission and absorption of gamma rays. This effect is normally studied in solids but has been observed in some unique liquids. The liquid crystalline state, in particular the smectic phase, presents a completely different environment in which to study this effect. In the smectic C phase the molecules are bound within planes with the spacing between the planes being less than the length of the molecule; i.e., the molecules are tilted in the layers. The motion of the molecules, on the other hand, is more fluid-like within the planes. This property is characterized by a rather large anisotropy in the viscosity.

Uhrich chose the compound 4,4'-di-*n*-heptyloxyazoxybenzene for this study primarily because it not only shows the smectic phase but also the nematic phase at a higher temperature. This combination is necessary to arrange the smectic liquid into a "single crystal." In the nematic phase the molecules were aligned by a magnetic field; then, while aligned, the temperature was lowered into the smectic phase.

The smectic phase obtained in this manner retains the preferred direction of orientation of the molecules characteristic of that in the nematic in a magnetic field. The choice of the solute was found to be more difficult in that it was necessary to have an iron containing molecule which (a) would not easily react with the solvent, (b) would order in the liquid crystal, and (c) would display the quadrupole splitting.

Mössbauer spectra for 1,1'-diacetylferrocene ordered by the smectic phase of 4,4'-di-*n*-heptyloxyazoxybenzene are shown in Figure 42. The splittings between the lines are determined by the value of the quadrupole coupling constant. The sum of the areas under the peaks is proportional to the recoil-free fraction. This quantity was observed to decrease sharply to

FIGURE 42

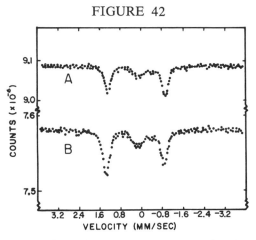

Experimental Mössbauer spectrum of ^{57}Fe in 1,1'-diacetylferrocene dissolved in 4,4'-di-*n*-heptyloxyazoxybenzene at 75°C in the smectic liquid crystalline state. The outer pair of lines is from the ordered solute. The smaller central line is due to the Fe in the sample container.[232]

a non-zero value as the sample was heated from the solid state into the smectic phase, then remain relatively constant as the temperature continued to rise through the smectic range. As the smectic-nematic transition was approached, it would then rapidly decrease to zero. The most interesting quantity is the ratio of the areas of each peak. This quantity depends upon the direction the incident gamma ray makes with the direction of the electric field gradient. This angle can be easily varied by changing the direction of the magnetic field relative to the direction of the incident gamma ray beam. Figure 43 shows the variation of the area ratio with angle which further shows that the preferred direction of the solute molecules is uniform throughout the sample.

The implications of this experiment are that it is possible to study the structure of the smectic phase by measurements of recoil-free fraction and to make Mössbauer studies of ordered molecules which do not form single crystals in the solid state.

X. OTHER PHYSICAL PROPERTIES

A. Viscosity

The early measurements[233-236] of viscosity in a nematic liquid crystal showed that this quantity was anisotropic as one might expect. The value of viscosity for p-azoxyanisole was measured[235] to be 2.4 cp for the case where the molecules were oriented parallel to the direction of flow while perpendicular orientation at the same temperature gave a value of 9.2 cp. These values vary little from those measured in the isotropic liquid,[237] namely 2.9 cp at the temperature a few degrees above the transition. The temperature dependence[233] of viscosity was also measured for p-azoxyanisole and found to be quite unusual. Its value would decrease with increasing temperature until the temperature approached the vicinity of the nematic-isotropic transition where it would then increase until the transition was reached.

In recent years the literature on viscosity measurements in nematic liquids has been meager although some related theoretical work has appeared.[63-66] Many of the conventional methods used for measuring viscosity in normal liquids cannot be used for liquid crystals because of surface orientation effects. Surface effects have been clearly demonstrated by Fisher and Fredrickson.[237] They used capillary tube viscometers, each with different diameters. The interior surfaces of the capillaries were pretreated in two different ways in order to obtain parallel or perpendicular orientation of the molecules at the surface. For perpendicular alignment the surfaces were washed with sulfuric acid-dichromate, rinsed in distilled water and dried. For parallel alignment hair was drawn through the tube numerous times prior to filling with the liquid crystal. The results of their measurements for perpendicular orientation are shown in Figure 44 for four different

FIGURE 43

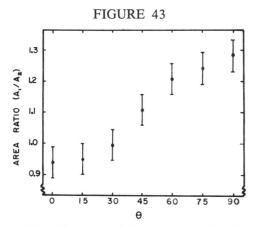

Plot of the area ratio for the two quadrupole split lines of 1,1′-diacetylferrocene in 4,4′-di-n-heptyloxyazoxybenzene in the smectic C phase as a function of the angle θ between the aligning magnetic field and the gamma-ray beam.[232]

FIGURE 44

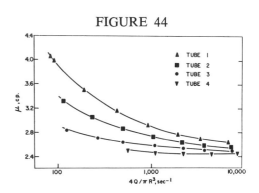

Viscosity behavior of p-azoxyanisole in the nematic phase at 121.8°C with the surface of the capillary treated so as to produce a perpendicular orientation of the molecules at the surface.[237] With permission.

tube diameters and for varying flow rates Q. The apparent viscosity, μ, was calculated from the Hagen-Poiseuille equation:

$$\mu = \frac{PR/2}{4Q/\pi R^2} \qquad (43)$$

where R is the radius of the capillary, P the pressure gradient. For Newtonian fluids in the absence of wall effects, μ is a constant function of $4Q/\pi R^3$. This equation was found to hold true in the isotropic phase of p-azoxyanisole but was found in considerable error in the nematic phase as illustrated in Figure 44. For surfaces prepared for parallel alignment, the viscosity more closely follows the Hagen-Poiseville equation. In both cases the viscosity approaches a common limit for high flow rates and large tube diameters. For perpendicular alignment they were able to estimate depth, δ, into the capillary to which this surface orientation affected the viscosity. Their values varied from 8 to 1 micron for slow to fast flow rates, respectively. Aside from the anisotropy, it appears that nematic liquids flow much like isotropic liquids far from the boundary of the capillary.

The flow of cholesterics and smectics is considerably more complicated. Viscosity in these systems has been studied in detail by Porter et al.[238-240] and, more recently, by Helfrich.[241] Capillary flow has been described as being more like a plug sliding down the tube as opposed to a liquid where the velocity of flow varies with distance from the boundary of the capillary.

B. Ultrasonics

The absorption and velocity of longitudinal sound waves have been measured in several thermotropic liquid crystal systems[242-245] and recently in lyotropic systems.[246] These studies have been primarily concerned with the absorption and velocity at the transitions. They all report that the absorption passes through a maximum at the isotropic-liquid crystal transition. The velocity appears to change discontinuously at the transition, passing through a minimum at that point, in the case of thermotropic systems. These results are similar to those seen near the critical point of the vapor-liquid transition in rare gases.[247] Based upon the same theories used for absorption near the critical point in rare gases, Edmonds and Orr[247] have applied the theory of Maier and Saupe to explain absorption in the nematic phase near the isotropic transition. They find that reasonable agreement between theory and experiment can be obtained by a combination of these theories.

The absorption and velocity effects seen at the isotropic-liquid crystal transition do not appear to be present at phase transitions which occur within the liquid crystal phase, such as the nematic-smectic[245] and cholesteric-smectic.[248]

C. Brillouin Scattering

Brillouin scattering has been studied in liquid crystals. The Doppler-shifted components of a laser beam have been observed[249,250] in the liquid crystal phase of cholesteryl 2-(2-ethoxyethoxy)ethyl carbonate. Dispersion of the hypersonic waves was observed by Durand et al.[250] near the clearing point.

D. Positron Annihilation

Positron lifetimes in cholesteryl propionate have been studied by Cole, Merrit and Walker.[251,252] The effect of ordering on the distribution of annihilation quanta from p-azoxyanisole has been reported.[253]

E. Birefringence

A theory of the birefringence of nematic liquid crystals has been developed by Chandrasekhar, Krishnamurti and Madhusudana.[254] The authors make an extensive calculation taking account of several terms in the intermolecular potential energy including dipole-dipole, anisotropic dispersion, induction and repulsion interactions. The theory explains the experimentally observed result that the temperature coefficient of the extraordinary index is large and negative, whereas the ordinary index is small and positive. It is shown that dispersion and repulsion forces play a dominant role in determining the temperature variation of the birefringence.

XI. LYOTROPIC LIQUID CRYSTALS

It seems in order in a review such as this one to include a brief section on the structure of lyotropic liquid crystals. It is the authors' opinion that some of the most significant future developments in liquid crystals will be in this area. By the combination of two or more components, liquid crystalline systems in great numbers can be prepared. These systems occur in both the inanimate and animate world.

In inanimate systems, one finds lipid-water systems very common. Three excellent reviews of the field are available.[255 256 262]

The role of liquid crystals in living systems appears to be a very important one. Slight changes in composition and in physical and chemical properties can materially affect the formation, continuation, or cessation of the liquid crystalline state, a delicate balance characteristic also of many biological processes. Catalytic processes in biological systems could readily find a favorable environment in the liquid crystalline structure. We shall not pursue this aspect of liquid crystals further but will cite two pertinent references written by Stewart.[257 258]

A. Amphiphilic Compounds (Amphiphiles)

In this discussion of lyotropic liquid crystals we shall concern ourselves with the structure of amphiphilic compounds. Amphiphilic compounds are characterized by having in the same molecule two groups which show greatly different solubility properties. These groups are (1) a hydrophilic portion of the molecule which tends to be water soluble and insoluble in hydrocarbons and (2) a lipophilic group which tends to be water insoluble and soluble in hydrocarbon. Depending on the relative contribution of each of the molecular portions, amphiphilic compounds may range from essentially hydrophilic (water soluble) to predominantly lipophilic (water insoluble). The amphiphilic molecules which have the greatest tendency to form liquid crystalline systems with water are those in which the hydrophilic and lipophilic units are strong and rather equally matched. In systems which form lyotropic liquid crystals, order arises as a consequence of selective interactions among two or more molecular species. Typical hydrophilic groups are $-OH$, $-O(CH_2-CH_2-O)_nH$, $-CO_2H$, $-CO_2Na$, $-SO_3K$, $-NMe_3Br$, $-PO_4-CH_2-CH_2-NH_3^+$ and typical lipophilic groups are $-C_nH_{2n+1}$, $C_nH_{2n+1}-\langle\rangle-$, $C_nH_{2n+1}-O_2C-CH-CH_2CO_2-C_nH_{2n+1}$.

Amphiphiles have, as a consequence of their dual character, striking solubility characteristics. Since they show miscibility in both water and organic solvents, they show marked co-solvent effects. Rather dilute soap solutions can dissolve a variety of organic compounds which would not dissolve in any significant amount in water alone. An amphiphile (e.g., Aerosol OT) dissolved in a hydrocarbon can solubilize water or organic compounds which are not normally soluble in the hydrocarbon. There are various ways to interpret this "solubilization." A reasonable one is that the amphiphile acts as the co-solvent for water and the hydrocarbon.

Binary systems of amphiphiles can be divided into binary aqueous amphiphile solutions and binary solutions of amphphiles in organic solvents. Sub-classes of binary aqueous solutions involve ionic amphiphiles (e.g., salts of fatty acid) and nonionic amphiphiles (e.g., poly (ethylene glycol) derivatives). Systems involving binary solutions of amphiphiles in organic solvents can be sub-divided into soaps of high crystallinity (e.g., soaps which have been heated) and solutions of amphiphiles of low crystallizing tendency (e.g., alkali dinonylnaphthalene sulfonates in a variety of organic solvents). All of these have been described by Winsor.[256]

B. Micelles

In solutions of amphiphiles there is a tendency for like groups in the molecules to associate. In aqueous solutions of salts of aliphatic acids, the non-polar portions of the paraffin chains tend to associate in the system, leaving the polar portions of the molecules to associate with the water molecules which surround them. The essentially transitory packing of amphiphile molecules arising in this manner is termed *micelles*. Micelles can be considered as following a statistical distribution of the molecules, and one should be careful not to consider them as entities with well-defined geometric shapes. In the liquid crystalline state the micellar groupings show marked persistence and long-range intermicellar order. Changes in conditions such as temperature, mechanical stress or concentration of components will immediately alter the order.

The two main types of intermolecular forces in amphiphilic solutions are electrostatic and dispersive. The electrostatic interactions (hydrophilic groups) arise from charges on ions and from dipoles. Interactions between ions, dipoles, and ions and dipoles lead to attractive orientations. Thermal motion tends to disrupt these arrangements. The dispersive forces which are attributed to movements are generally weak, but in certain cases they do predominate in effecting molecular arrangements.

C. Classes of Lyotropic Structures

A systematic classification of lyotropic liquid crystals has become possible primarily as a result of the X-ray studies of Luzzati, Skoulios and their co-workers.[259-261] From their studies we can now describe the various lyotropic mesophase structures which have been reported in the literature. These structures are characterized by their optical and structural properties (determined primarily by X-ray methods). Schematic representations of the different types of molecular packing are presented in Figures 45 and 46. The symbolisms used by different authors to identify a class are generally unique to the authors so we are selecting one of our own which is undoubtedly no better than others which have been proposed. The systems which have molecular packings to give lamellar structures we shall represent with the symbol L with an appropriate subscript to distinguish one molecular arrangement from another. Those with the cubic packing will be identified with the symbol C and those with particle structure with the symbol P. Subscripts will be used in the last cases to distinguish different molecular arrangements within a major class. The summary of this classification and a brief description of the properties of each are presented in Table 9.

There are two ways in which water-containing mesophases can form. The particles which

FIGURE 45a

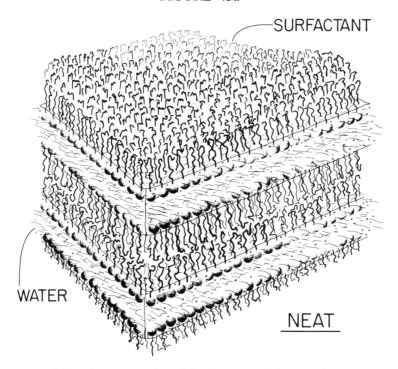

Schematic representation of the structure of the neat phase.

FIGURE 45b

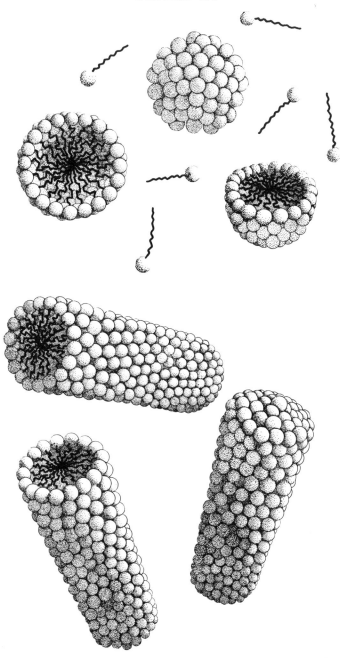

Schematic representation of rod-like and spherical particles. (Figures 45 a and b reproduced by permission of *J. Soc. Cosmet. Chem.*)

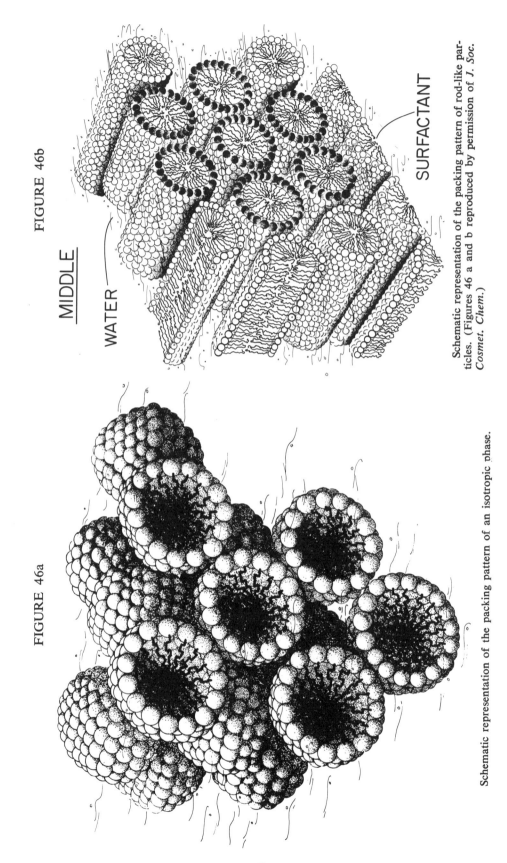

FIGURE 46b

MIDDLE

WATER

SURFACTANT

Schematic representation of the packing pattern of rod-like particles. (Figures 46 a and b reproduced by permission of *J. Soc. Cosmet. Chem.*)

FIGURE 46a

Schematic representation of the packing pattern of an isotropic phase.

TABLE 9

Different Types of Lyotropic Liquid Crystalline Phases*

I. Structural arrangement displaying Bragg spacing ratio $1:\frac{1}{2}:\frac{1}{3}$ with "linear" symmetry.

Class	Description	Common Notation in Literature
L_1	Lamellar packing with coherent double layers of molecules and ions separated by water. *Neat phase type.* (See Figure $47L_1$)	Neat Phase
L_2	Lamellar packing with coherent single layers of molecules and ions separated by water. *Single-layered lamellar type.*	
L_3	Lamellar packing with coherent double layers of molecules and ions separated by water. *Mucous woven type.* (See Figure $47L_2$)	

II. Lyotropic liquid crystals with Particle structure displaying Bragg spacing ratio $1:\frac{1}{2}:\frac{1}{3}:\frac{1}{4}$.

P_1	Rod-like particles with organic core surrounded with water. Rods with predominantly quadratic cross-section in tetragonal arrangement. *Normal two-dimensional tetragonal type.* (See Figure $48P_1$)	White Phase
P_2	Rod-like particles with water core in organic environment. Rods with predominantly quadratic cross-section in tetragonal arrangement. *Reversed two-dimensional tetragonal type.* (See Figure $48P_2$)	
P_3	Rod-like particles with organic core in aqueous environment. Rods with rectangular cross-section in an orthorhombic array. *Normal two-dimensional rectangular type.* (See Figure $48P_3$)	Rectangular Phase

III. Lyotropic liquid crystals displaying Bragg spacing ratio $1:\dfrac{1}{\sqrt{3}}:\dfrac{1}{\sqrt{4}}:\dfrac{1}{\sqrt{7}}$. Particle structure with molecules arranged in two-dimensional hexagonal symmetry.

P_{H-1}	Rod-like particles with organic core in aqueous environment. Cylindrical to hexagonal cross-section in hexagonal array. (1) *Middle phase type;* (2) *Normal two-dimensional hexagonal type* (see Figure $49P_{H-1}$.)	(1) Middle Phase (2) Hexagonal Phase-I
P_{H-2}	Rod-like particles with aqueous core in organic environment. Cylindrical to hexagonal cross-section in hexagonal array. *Reversed two-dimensional hexagonal type.* (See Figure $49P_{H-2}$)	Hexagonal Phase-II
P_{H-3}	Rod-like particles with complex structure in aqueous environment. *Complex two-dimensional hexagonal type.*	Complex Hexagonal Phase

* Table prepared by modifying Table 1 after P. Ekwall et al.[262]

TABLE 9 (Continued)

IV. Lyotropic liquid crystals displaying cubic symmetry.

  IV-1. Isotropic lyotropic liquid crystals with spherical to dodecahedral particles arranged in face-centered cubic lattice.

Class	Description	Common Notation in Literature
C_{f-1}	Particles with organic core in aqueous environment. *Normal face-centered cubic type.* (See Figure 50C_{f-1})	Cubic Phase, C_{I-1}
C_{f-2}	Particles with water core in organic environment. *Reversed face-centered cubic type.* (See Figure 50C_{f-2})	Cubic Phase, C_{I-2}
C_{f-3}	Particles with complex structure. *Complex face-centered cubic type.*	

  IV-2. Isotropic lyotropic liquid crystals with spherical particles packed in body-centered cubic lattice

C_{b-1}	Particles with organic core in aqueous environment. *Normal body-centered cubic type.*	
C_{b-2}	Particles with complex structure. *Complex body-centered cubic type.*	

generate the mesophases can be built around a core of hydrophilic groups, usually hydrated, or around a water core. The organic portion of the molecule is projected outward. This structure is known as the reversed type. The "normal" type has the organic portions of the molecules projected toward the center of the rod-like particle and the hydrophilic group is directed outward and surrounded by water.

We shall give schematic drawings (Figures 47–50) of several of the structures described in Table 9. The drawings are from Ekwall, Mandell and Fontell[262] who, in turn, based their drawings to a great extent on their own work and that of Luzzati, Skoulios et al.[259-261] Even though the arrangement of the molecular aggregates is quite well known, we have little reliable information on the arrangement of the molecules within the aggregates. We know that the hydrophilic groups in the amphiphilic molecule are soluble in water and are thus fixed in the water-amphiphile interfaces. It is, therefore, fairly safe to assume that the organic portion of the molecules has an average principal orientation which can vary from phase to phase.

Schematic Drawings of Molecular Arrangements in Lamellar Structures

FIGURE 47L_1

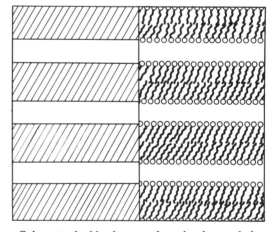

Coherent double layers of molecules and ions separated by water. Proposed neat structure with tilted layers. With permission.

FIGURE 47L₃

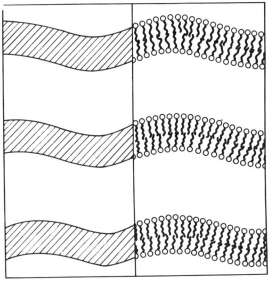

Coherent double layers of molecules and ions separated by water. Proposed structure for mucous woven type with molecules arranged vertically.[262] With permission.

FIGURE 48P₂

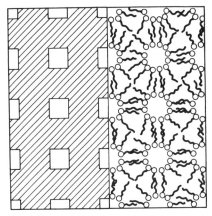

Rod-like particles with water core in organic environment. Cylinders with predominantly quadratic cross section in tetragonal arrangement. Reversed two-dimensional tetragonal type. With permission.

Schematic Drawings of Molecular Arrangements in Tetragonal and Orthorhombic Structures

FIGURE 48P₁

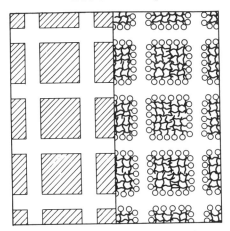

Rod-like particles with organic core surrounded with water. Cylinders with predominantly quadratic cross section in tetragonal arrangement. Normal two-dimensional tetragonal type. With permission.

FIGURE 48P₃

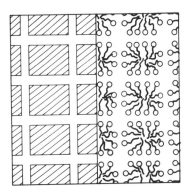

Rod like particles with organic core in aqueous environment. Cylinders with rectangular cross section in an orthorhombic array. Normal two-dimensional rectangular type.[262] With permission.

Schematic Drawings of Molecular Arrangements of Two-Dimensional, Hexagonal Liquid Crystalline Structures

FIGURE 49P_{H-1}

Rod-like particles with organic core in aqueous environment. Cross section in hexagonal array. Proposed for middle phase. With permission.

FIGURE 49P_{H-2}

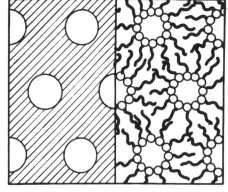

Rod-like particles with aqueous core in organic environment. Cross section in hexagonal array. Proposed structure for reversed two-dimensional hexagonal type. With permission.

Schematic Drawings of Molecular Arrangements in Isotropic Liquid Crystals

FIGURE 50C_{f-1}

Particles with organic core in aqueous environment. Proposed structure for normal face-centered cubic type. With permission.

FIGURE 50C_{f-2}

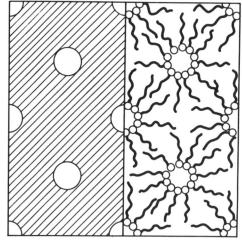

Particles with water core in organic environment. Proposed structure for reversed face-centered cubic type.[262] With permission.

D. Composition of a Typical System and Occurrence of Various Liquid Crystalline Phases in that System

Extensive phase studies of two- and multicomponent systems have been made by McBain[263] and his co-workers and Ekwall and his co-workers.[262] Contributions from Ekwall's laboratories are summarized in Reference 262. The phase diagrams for different soaps are basically similar. For our example, we have selected the phase diagram for sodium laurate[263] and water (Figure 51). A few remarks about this phase diagram seem in order. The T_i line indicates the variation in the temperature of breakdown of liquid crystalline order. The T_c line may be considered as representing the depression of the melting point of the laurate by water. For temperatures higher than those identified by this line, the lattice of the solid soap is entirely broken down and the systems are liquid crystalline or liquid. The structure of the neat and middle phases are described in Table 9. Excellent discussion of phase diagrams of other binary and ternary systems may be found in articles by Ekwall, Mandell and Fontell[262] and Lawrence.[255]

Studying the phase diagram in Figure 51, you will notice two darkened portions on the curve. These areas represent compositions where the system takes on an isotropic structure. This phase has the molecular arrangement proposed in Figures 46(a) and 50; it is isotropic and has a very high viscosity. The compositions of the systems between the neat and middle phases are undoubtedly conjugate mixtures. The same can undoubtedly be said about intermediate phases along the lines that tie other phases together. The intermediate phases under a polarizing microscope do not have well-defined textures as one finds within the region of a phase.

A schematic diagram integrating the interrelationships of lyotropic liquid crystals cannot portray the large volume of results which has been accumulated on these systems, but it provides an impression of the state of the art. In Figure 52 we have attempted to very briefly describe properties which are characteristic of lyotropic systems. Note that variation of the ratio of amphiphile to water can alter the structure from crystalline to a true solution. The intermediate structures may be lamellar, cubic, hexagonal, complex hexagonal and/or micellar.

The reader is referred to Ekwall, Mandell and Fontell[262] and to Winsor[256] for a more complete schematic presentation of the interrelationships among homogeneous solutions, micellar systems and liquid crystalline systems.

E. Transitions Between Different Lyotropic Liquid Crystal Phases

The lamellar structure is intermediate between the normal and reversed structures in lyotropic liquid crystals. By varying the concentration of the components in a system, various structures follow each other in a certain sequence. In Figure 53 we have attempted to show some of these transitions in a schematic way. The outline follows that proposed by Ekwall et al.[262] The lettering system found in the figure follows the scheme given in Table 9; the symbols S_1 and S_2 refer to solutions.

FIGURE 51

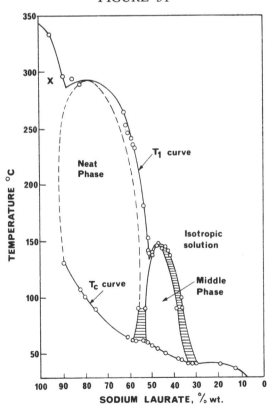

Phase diagram for the sodium-laurate-water system.[255]

FIGURE 52

Suggested Structural Arrangement	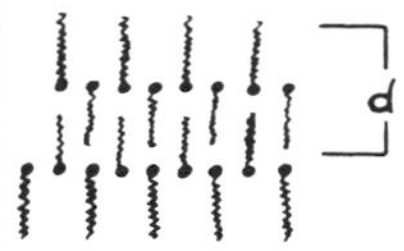		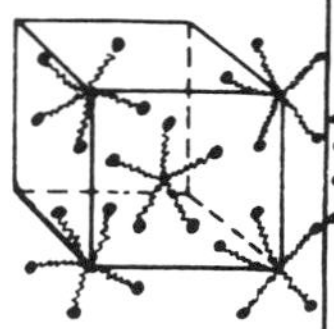	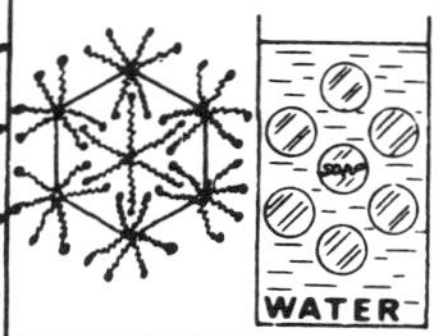		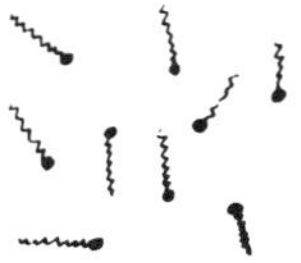
% Water* (approximate range)	0	5-22 -- 50	23-40	34-80	30-99.9	Greater than 99.9
Physical State	Crystalline	Liquid crystalline, lamellar	Liquid crystalline, face-centered cubic	Liquid crystalline, hexagonal compact	Micellar solution	Solution
Gross Character	Opaque solid	Clear, fluid, moderately viscous	Clear, brittle, very viscous	Clear, viscous	Clear, fluid	Clear, fluid
Freedom of Movement	None	2 directions	Possibly none	1 direction	No restrictions	No restrictions
Microscopic Properties (crossed nicols)	Birefringent	Neat soap texture	Isotropic with angular bubbles	Middle soap texture	Isotropic with round bubbles	Isotropic
X-ray data	Ring pattern 3-6 Å	Diffuse halo at about 4.5 Å	Diffuse halo at about 4.5 Å	Diffuse halo at about 4.5 Å		
Structural Order	3 dimensions	1 dimension	3 dimensions	2 dimensions	None	None

*The different percentages of water show that different amphiphiles require different amounts of water. For soaps, the lamellar structure generally occurs between 5-22% water; with some lipophiles the water may be as high as 50%. The cubic structure generally occurs between 23-40%.

Some properties of lyotropic systems composed of an amphiphile and water.

FIGURE 53

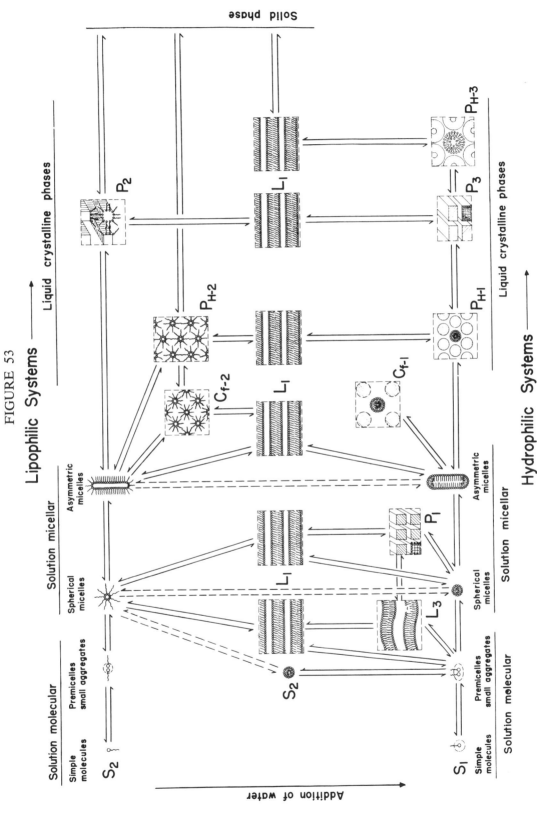

Schematic representation of the transitions between different states of lipophilic and hydrophilic systems. (Modified from Ekwall et al.[262] With permission.

The transformation between liquid crystalline phases is represented by solid lines and those relating reverse and normal structures in homogeneous phases are given by dotted lines.

At the bottom of Figure 53 there are successive transformations from solution, to micelles, to liquid crystals. The liquid crystalline structures are of the normal type, i.e. an organic core in a water continuum. At the top of the figure there is a comparable relationship in a lipophilic environment as one moves from left to right. The systems at the top of the graph are of the reverse type, i.e. a core of hydrophilic groups. On moving from top to bottom in the figure, the amount of water increases while the amphiphile concentration decreases. It can be seen that many of the liquid crystalline phase changes also take place as the water to amphiphile ratio varies.

An interesting relationship given in Figure 53 is the central location of the lamellar phase. Evidence accumulated to date indicates the transition from normal to a reversed liquid crystalline phase takes place via the lamellar phase. In the micellar solutions a transformation from the normal to the reversed type can take place (represented by dotted lines), but it is not known whether the intermediate lamellar structure is involved.

Even though the schematic presentation in Figure 53 cannot portray an exact picture of phase relationships in lyotropic systems, it is a reasonable expression of the present state of our knowledge of the field. The scheme is applicable to binary and ternary systems.

F. Summary

A number of factors govern the stability of the various lyotropic structures.[262] These include: (1) ratio of water to organic substances; (2) molecular structure of the organic substances (e.g., relative bulkiness of the polar groups and the paraffinic units in the amphiphile molecule); and (3) packing density in the amphiphile parts of the mesophases as a whole and in their boundary layer.

Ekwall et al.[262] report that no transition from normal to reversed types of lyotropic systems, or vice versa, has been observed to this date in binary systems of an amphiphile and water, but the process occurs readily in ternary systems composed of two amphiphiles and water. The transition seems to be closely related to the ratio of water to amphiphile, the change to the reversed type being promoted by the reduction of water. In paraffinic compounds, where both normal and reversed types exist, the normal structures occur with all the water bound or with part completely free while the reversed type occurs if all or nearly all the water is attached to the hydrophilic group.

As pointed out previously, the lamellar structure serves as a link between the reversed and normal structures. The lamellar structure has the capacity to exist over rather wide ranges of amphiphile-to-water ratios. The other liquid crystalline structures, however, exist within rather restricted conditions of amphiphile-to-water ratios.

Acknowledgments

The authors wish to express their thanks to Professor A. Saupe for helpful advice during the preparation of the manuscript. The help of Dr. J. T. S. Andrews on the thermodynamics section is gratefully acknowledged. Dr. Derry Fishel was very helpful on matters of nomenclature and his contributions are gratefully recognized. Dr. E. Gelerinter read the section on electron paramagnetic resonance and offered helpful suggestions.

It is with much gratitude that we thank Professor H. Sackmann for permission to present in Table 3 examples of class c_2 and e_3. This permission was granted prior to Professor Sackmann's own publication of the compounds.

Many publishers generously granted permission to reproduce figures from scientific literature. With each figure, where proper, we have identified the reference from which the figure was taken. We wish to express our thanks to the publishers of the following journals and books for granting us permission to reproduce material from their publications: *Journal of the American Chemical Society, Molecular Crystals and Liquid Crystals, Journal of the Society of Cosmetic Chemists, Transactions of the Faraday Society, Discussions of the Faraday Society, Molecular Physics, Journal of the Physical Society of Japan, The Journal of Chemical Physics, Physical Review Letters,*

Journal of Magnetic Resonance, Advances in Chemistry Series, Fortschritte der Chemischer Forschung, Zeitschrift für Chemie, Angewandte Chemie, Zeitschrift für Naturforschung, Liquid Crystals and Ordered Fluids (Plenum Press), and Liquid Crystals 2 (Gordon and Breach Science Publishers).

REFERENCES

1. Brown, G.H., *Chemistry*, 40, 10 1967.
2. Brown, G.H., *Anal. Chem.*, 41, 26A 1969.
3. Brown, G.H. and Shaw, W.G., *Chem. Rev.*, 57, 1049 1957.
4. Chistyakov, I.G., *Soviet Phys. Cryst.*, 5, 917 1961.
5. Usol'tseva, V.A. and Chistyakov, I.G., *Russ. Chem. Rev.*, 32, 495, 1963.
6. Saupe, A., *Angew. Chem. Intern. Ed.* (English), 7, 97 1968.
7. Sackmann, H. and Demus, D., *Fortschr. Chem. Forsch.*, 12 (2), 349 1969.
8. Gray, G.W., *Molecular Structure and the Properties of Liquid Crystals,* Academic Press, New York, 1962.
9. Liquid Crystals, *Proc. 1st Int. Liquid Crystal Conf.*, coordinated by G.H. Brown, G.J. Dienes, and M.M. Labes, Gordon and Breach Science Publishers, Inc., New York, 1967.
10. Liquid Crystals 2, *Proc. 2nd Int. Liquid Crystal Conf.*, Part I, Brown, G.H., Ed., Gordon and Breach Science Publishers, Inc., New York, 1969.
11. Liquid Crystals 2, *Proc. 2nd Int. Liquid Crystal Conf.*, Part II, Brown, G.H., Ed., Gordon and Breach Science Publishers, Inc., New York, 1969.
12. Lehmann, O., *Z. Phys. Chem.*, 4, 462 1889.
13. Kast, W., in Landolt-Börnstein, Vol. 2, part 2a, 6th ed., Springer, Berlin, 1969, 266.
14. Gray, G.W., *Mol. Cryst. Liquid Cryst.*, 7, 127 1969.
15. Gulrich, L.W. and Brown, G.H., *Mol. Cryst.*, 3, 493 1968.
16. Wiegand, C., *Z. Naturforsch.*, 9b, 516 1954.
17. Shubert, H., Eissfeldt, I., Lange, R., and Trefflich, F., *J. Prakt. Chem.*, (4) 33, 265 1966.
18. Vorländer, D., *Z. Phys. Chem.*, A126, 449 1927.
19. Vorländer, D., *Ber. Deut. Chem. Ges.*, 54, 2261 1921.
20. Ennulat, R.D., *Mol. Cryst. Liquid Cryst.*, 8, 247 1969.
21. Bennett, M.G. and Jones, B., *J. Chem. Soc.*, 1939, 420.

22. Bacon, W.E. and Brown, G.H., *Mol. Cryst. Liquid Cryst.*, 6, 155 1969.

23. Demus, D. and Sackmann, *Z. Phys. Chem.* (Leipzig), 222, 127 1963.

24. Wu, Y.C., Master's Thesis, Kent State University, August 1969.

25. Sackmann, H., Diele, S., and Brand, P., presented at the International Crystallographic Conference, Stony Brook, N.Y., August 1969.

26. Demus, D., Kunicke, G., Neelsen, J., and Sackmann, H., *Z. Naturforsch.*, 23a, 84 1968.

27. Demus, D., Sackmann, H., Kunicke, G., Pelzl, G., and Salffner, R., *Z. Naturforsch.*, 23a, 76 1968.

28. de Vries, A., *Mol. Cryst. Liquid Cryst.*, 10, 31 1970.

29. de Vries, A., *Acta Crystallogr.*, A25, S 135 1969.

30. Falgueirettes, J., *Bull. Soc. Fr. Minéral. Christallogr.*, 82, 172 1959.

31. Luzzati, V., Mustacchi, H., Skoulios, A., and Husson, F., *Acta Crystallogr.*, 13, 660 1960.

32. (a) Chistyakov, I.G. and Vainshtein, B.K., *Soviet Phys. Cryst.*, 8, 458 1964;
    (b) Chistyakov, I.G., et al., *Liquid Crystals 2,* Part II, Brown, G.H., Ed., Gordon and Breach Science Publishers, Inc., New York, 1969, 813.

33. Chaikovskii, V.M. and Chistyakov, I.G., *Soviet Phys. Cryst.*, 13, 229 1968.

34. Chistyakov, I.G. and Chaikovskii, V.M., *Soviet Phys. Cryst.*, 12, 770 1968.

35. Kosterin, E.A. and Chistyakov, I.G., *Soviet Phys. Cryst.*, 13, 229 1968.

36. Vainshtein, B.K., Chistyakov, I.G., Kosterin, E.A., and Chaikovskii, V.M., *Mol. Cryst. Liquid Cryst.*, 8, 457 1969.

37. Chistyakov, I.G. and Chaikovskii, V.M., *Mol. Cryst. Liquid Cryst.*, 7, 269 1969.

38. Lehmann, O., *Flussige Kristalle,* Leipzig, 1904.

39. Bose, E., *Physik. Z.*, 8, 513 1907; 10, 230 1909.

40. Ornstein, L.S., *Z. Kristallogr.*, 79, 10 1931.

41. Ornstein, L.S. and Zernike, F., *Physik. Z.*, 19, 134 1918.

42. Ornstein, L.S. and Kast, W., *Trans. Faraday Soc.*, 29, 881 1933.

43. Zocher, H., *Physik. Z.*, 28, 790 1927.

44. Zocher, H., *Trans. Faraday Soc.*, 29, 931 1933; 29, 945 1933.

45. Oseen, C.W., *Trans. Faraday Soc.*, 29, 883 1933.

46. Frank, F.C., *Discussions Faraday Soc.*, 25, 19 1958.

47. Ericksen, J.L., *Arch. Ration. Mech. Anal.*, 9, 371 1962.

48. Ericksen, J.L., *Arch. Ration. Mech. Anal.*, 10, 189 1962.

49. Ericksen, J.L., *Phys. Fluids,* 9, 1205 1966.

50. Ericksen, J.L., *Mol. Cryst. Liquid Cryst.*, 7, 153 1969.

51. Saupe, A., *Z. Naturforsch.*, 15a, 815 1960.

52. de Vries, H., *Acta Crystallogr.*, 4, 1951.

53. de Gennes, P.G., *Mol. Cryst. Liquid Cryst.*, 7, 325 1969.

54. Zocher, H., *Mol. Cryst. Liquid Cryst.*, 7, 165 1969.

55. Grandjean, F., *Compt. Rend.*, 172, 71 1921.

56. Cano, R., *Bull. Soc. Fr. Minéral. Cristallogr.*, 91, 20 1968.

57. Kassubek, P. and Meier, G., *Mol. Cryst Liquid Cryst.*, 8, 305 1969.

58. Orsay Liquid Crystal Group, *Phys. Lett.*, 28A, 687 1969.

59. Friedel, J. and Kleman, M., see reference 58.

60. Ericksen, J.L., *Arch. Ration. Mech. Anal.*, 4, 231 1969.

61. Ericksen, J.L., *Trans. Soc. Rheol.*, 5, 23 1961.

62. Ericksen, J.L., *Trans. Soc. Rheol.*, in press.

63. Ericksen, J.L., *Appl. Mech. Rev.*, 20, 1029 1967.

64. Ericksen, J.L., *J. Fluid Mech.*, 27, 59 1967.

65. Leslie, F.M., *Quart. J. Mech. Appl. Math.*, 19, 357 1966.

66. Leslie, F.M., *Arch. Ration. Mech. Anal.*, 28, 265 1968.

67. Leslie, F.M., *Proc. Roy. Soc.* (London), A307, 359 1968.

68. Coleman, B.D., *Arch. Ration. Mech. Anal.*, 20, 41 1965.

69. Wang, C.C., *Arch. Ration. Mech. Anal.*, 20, 1 1965.

70. Leslie, F.M., *Mol. Cryst. Liquid Cryst.*, 7, 407 1969.

71. Porter, R.S. and Johnson, J.F., *J. Phys. Chem.*, 66, 1826 1962.

72. Porter, R.S., Barrall, E.M., and Johnson, J.F., *J. Chem. Phys.*, 45, 1452 1966.

73. Chandrasekhar, S., *Proc. Roy. Soc.* (London), 281, 92 1964.

74. Saupe, A., *Mol. Cryst. Liquid Cryst.*, 7 59 1969.

75. Friedel, M.G., *Ann. Phys.*, 18, 273 1922.

76. Sackmann, H. and Demus, D., *Mol. Cryst.*, 2, 81 1966.

77. Arora, S.L., Taylor, T.R., Fergason, J.L., and Saupe, A., *J. Amer. Chem. Soc.*, 91, 3671 1969.

78. Arora, S.L., Taylor, T.R., and Fergason, J.L., *J. Org. Chem.*, in press.

79. Taylor, T.R., Fergason, J.L., and Arora, S.L., *Phys. Rev. Lett.*, 24, 354 1970.

80. Arora, S.L., Fergason, J.L., and Saupe, A., *Liquid Crystals* 2, Part II, Brown, G.H., Ed., Gordon and Breach Science Publishers, Inc., New York, 1969, 563.

81. Lehmann, O., *Flüssige Kristalle die Theorien des Lebens,* Engelmann, Leipzig 1906.

82. Zocher, H., *Mol. Cryst., Liquid Cryst.*, 7, 177 1969.

83. Saupe, A., Semi-annual Report No. 6, Contract No. F44620-67-C-0103, Air Force Office of Scientific Research, Washington, D.C.

84. Rapini, A., Papoular, M., and Pincus, P., *Compt. Rend. B.*, 267, 1230 1968.

85. Fréedericksz, V. and Zolina, V., *Zh. RF. Kharkov*, 59, 183 1927; *Trans. Faraday Soc.*, 29, 919 1933.

86. Massen, C.H. and Poulis, J.A., in *Ordered Fluids and Liquid Crystals,* Advances in Chemistry Series, No. 63, Porter, R.S. and Johnson, J.F., Eds., American Chemical Society, Washington, D.C., 1967, Chapter 7.

87. Neff, V.D., Gulrich, L., and Brown, G.H., *Mol. Cryst.*, 1, 225 1966.

88. Maier, W. and Meier, G., *Z. Naturforsch.*, 16A, 470 1961.

89. Maier, W. and Meier, G., *Z. Naturforsch.*, 16A, 262 1961.

90. Maier, W. and Meier, G., *Z. Naturforsch.*, 16A, 1200 1961.

91. Meier, G. and Saupe, A., *Mol. Cryst.*, 1, 4 1966.

92. Axmann, A., *Z. Naturforsch.*, 21A, 290 1966.

93. Weise, H. and Axmann, A., *Z. Naturforsch*, 21A, 1316 1966.

94. Axmann, A., *Mol. Cryst.*, 3, 471 1968.

95. Carr, E.G., *J. Chem. Phys.*, 38, 1536 1963.

96. Carr, E.G., *J. Chem. Phys.*, 39, 1979 1963.

97. Carr, E.F., *J. Chem. Phys.*, 42, 738 1965.

98. Carr, E.F., *J. Chem. Phys.*, 43, 3905 1965.

99. Twitchell, R.P. and Carr, E.G., *J. Chem. Phys.*, 46, 2765 1967.

100. Carr, E.F., Hoar, E.A. and MacDonald, W.T., *Bull. Amer. Phys. Soc.*, 13, 222 1968.

101. Carr, E.G., in *Ordered Fluids and Liquid Crystals,* Advances in Chemistry Series, No. 63, Porter R.S. and Johnson J.F. Ed., American Chemical Society, Washington, D.C., 1967, Chapter 8.

102. Carr, E.F., *Mol. Cryst. Liquid Cryst.*, 7, 253 1969.

103. Williams, R., *J. Chem. Phys.*, 39, 384 1963.

104. Kapustin, A.P. and Vistin, L.K., *Kristallografiya,* 10, 118 1965.

105. Elliot, G. and Gibson, J., *Nature,* 205, 995 1965.

106. Williams, R., in *Ordered Fluids and Liquid Crystals,* Advances in Chemistry Series, No. 63, Porter R.S. and Johnson J.F., Eds., American Chemical Society, Washington, D.C., 1967, Chapter 8.

107. Williams, R., *Nature,* 199, 273 1963.

108. Williams, R. and Heilmeier, G.H., *J. Chem. Phys.*, 44, 638 1965.

109. Heilmeier, G.H., *J. Chem. Phys.*, 44, 644 1965.

110. Kessler, J.O., Longley-Cook, M., and Rasmussen, W.O., *Mol. Cryst. Liquid Cryst.*, 8, 327 1969.

111. Williams, R., *Phys. Rev. Lett.*, 21, 342 1968.

112. Williams, R., *J. Chem. Phys.*, 50, 1324 1969.

113. Helfrich, W., *Phys. Rev. Lett.*, 21, 1518 1968.

114. Heilmeier, G.H., Zanoni, L.A., and Barton, L.A., *Appl. Phys. Lett.*, 13, 46 1968.

115. Heilmeier, G.H. and Goldmacher, J.E., *Appl. Phys. Lett.*, 13, 132 1968.

116. Heilmeier, G.H. and Zanoni, L.A., *Appl. Phys. Lett.*, 13, 91 1968.

117. Meyer, R.B., *Phys. Rev. Lett.*, 22, 918 1969.

118. Jules, J.C., Picot, C., and Fredrickson, A.G., *Ind. Eng. Chem., Fundam.*, 7, 84 1968.

119. Davison, L., *Phys. Rev.*, 180, 232 1968.

120. Kusabayashi, S. and Labes, M.M., *Mol. Cryst. Liquid Cryst.*, 7, 395 1969.

121. Fergason, J.L., *Mol. Cryst.*, 1, 293 1966.

122. Fergason, J.L., Goldberg, N.N., and Nadalin, R.J., *Mol. Cryst.*, 1, 309 1966.

123. Keating, P.N., *Mol. Cryst. Liquid Cryst.*, 8, 315 1969.

124. de Gennes, P.G., *Solid State Commun.*, 6, 163 1968.

125. Meyer, R.B., *Appl. Phys. Lett.*, 14, 208 1969.

126. Durand, G., Leger, L., Rondelez, F., and Veyssie, E., *Phys. Rev. Lett.*, 22, 227 1969.

127. Sackmann, E., Meiboom, S., and Snyder, L.C., *J. Amer. Chem. Soc.*, 89, 5981 1967.

128. Leclerq, M., Billard, J., and Jacques, J., *Compt. Rend.*, 266C, 654 1968.

129. Sackmann, E., Meiboom, S., Snyder, L.C., Meixner, A.E., and Dietz, R.E., *J. Amer. Chem. Soc.*, 90, 3567 1968.

130. Adams, J.E., Haas, W., and Wysocki, J., *Phys. Rev. Lett.*, 22, 92 1969.

131. Adams, J.E., Haas, W., and Wysocki, J., *J. Chem. Phys.*, 50, 2458 1969.

132. Muller, J.H., *Z. Naturforsch.*, 20a, 849 1965.

133. Muller, J.H., *Mol. Cryst.*, 2, 167 1966.

134. Harper, W.J., *Mol. Cryst.*, 1, 325 1966.

135. Wysocki, J., Adams, J., and Haas, W., *Phys. Rev. Lett.*, 20, 1024 1968.

136. Wysocki, J., Adams, J., and Haas, W., *Mol. Cryst. Liquid Cryst.*, 8, 471 1969.

137. Vistin, L.K. and Kapustin, A.P., *Soviet Phys. Cryst.*, 13, 284 1968.

138. Brout, R., *Phase Transitions*, W.A. Benjamin, Inc., New York, 1965.

139. Maier, W. and Saupe, A., *Z. Naturforsch.*, 15A, 287 1960.

140. Maier, W. and Saupe, A., *Z. Naturforsch.*, 14A, 882 1959.

141. Vorlander, D., *Z. Phys. Chem.*, 105, 211 1923.

142. Saupe, A., *Z. Naturforsch.*, 15A, 810 1960.

143. Tsvetkov, V.N., *Acta Physiochemica USSR* 16, 132 1942.

144. Fowler, R. and Guggenheim, E.A., *Statistical Thermodynamics*, Cambridge University Press, 1939, Chapter 13.

145. Arnold, H., *Z. Chem.*, 4, 211 1964.

146. Cotter, M.A. and Martire, D.E., *Mol. Cryst. Liquid Cryst.*, 7, 295 1969.

147. Porter, R.S., Barrall, E.M., and Johnson, J. F., *Accounts Chem. Res.*, 2, 53 1969.

148. Johnson, J.F., Porter, R.S., and Barrall, E.M., *Mol. Cryst. Liquid Cryst.*, 8, 1 1969.

149. Barrall, E.M., Porter, R.S., and Johnson, J.F., *J. Phys. Chem.*, 68, 2810 1964.

150. Barrall, E.M., Porter, R.S., and Johnson, J.F., *J. Chromatogr.*, 21, 392 1966.

151. Barrall, E.M., Porter, R.S., and Johnson, J.F., *J. Phys. Chem.*, 70, 385 1966.

152. Barrall, E.M., Porter, R.S., and Johnson, J.F., *J. Phys. Chem.*, 71, 1224 1967.

153. Ennulat, R.D., *Mol. Cryst.*, 3, 405 1968.

154. Arnold, H., *Z. Physik. Chem.*, (Leipzig), 225, 45 1964.

155. Arnold, H., *Z. Physik. Chem.*, (Leipzig), 226, 146 1965.

156. Arnold, H., *Z. Physik. Chem.*, (Leipzig), 231, 407 1966.

157. Arnold, H., *Z. Physik. Chem.*, (Leipzig), 234, 401 1967.

158. Arnold, H., *Z. Physik. Chem.*, (Leipzig), 239, 283 1968.

159. Arnold, H., *Z. Physik. Chem.*, (Leipzig), 240, 177 1969.

160. Arnold, H., *Z. Physik. Chem.*, (Leipzig), 240, 185 1969.

161. Arnold, H., *Mol. Cryst.*, 2, 63 1966.

162. Sakevich, N.M., *Russ. J. Phys. Chem.* (English), 42, 11 1968.

163. Torgalkar, A. and Porter, R.S., *J. Chem. Phys.*, 48, 3897 1968.

164. Runyan, W.R. and Nolle, A.W., *J. Chem. Phys.*, 27, 1081 1957.

165. Porter, R.S. and Johnson, J.F., *J. Appl. Phys.*, 34, 51 1963.

166. Hoyer, W.A. and Nolle, A.W., *J. Chem. Phys.*, 24, 803 1956.

167. Tsvetkov, V.N. and Krozer, S.P., *Sov. Phys.-Tech. Phys.*, 3, 1340 1958.

168. Ferguson, A. and Kennedy, S.J., *Phil. Mag.*, 26, 41 1938.

169. Tsvetkov, V.N., *Acta Physiochemica USSR* 19, 86 1944.

170. Tsvetkov, V.N. and Ryumtsev, E.I., *Soviet Phys. Cryst.*, 13, 225 1968.

171. Maier, W. and Englert, G., *Z. Phys. Chem.*, (Neue Folge) 19, 168 1959.

172. Maier, W. and Englert, G., *Z. Electrochem.*, 64, 689 1960.

173. Koller, K., Lorenzen, K., and Schwab, G., *Z. Phys. Chem.*, (Neue Folge), 44, 101 1965.

174. Chatelain, P., *Acta Crystallogr.*, 1, 315 1948.

175. Chatelain, P., *Acta Crystallogr.*, 4, 453 1951.

176. Orsay Liquid Crystal Group, *J. Chem. Phys.*, 51, 816 1969.

177. de Gennes, P.G., *Compt. Rend.*, 266, 15 1968.

178. Doane, J.W. and Visintainer, J.J., *Phys. Rev. Letters*, 23, 1421 1969.

179. Adams, J., Haas, W., and Wysocki, *Mol. Cryst. Liquid Cryst.*, 8, 9 1969.

180. Cameron, L.M., *Mol. Cryst. Liquid Cryst.*, 7, 235 1969.

181. Kapusten, A.P., *Soviet Phys. Cryst.*, 12, 443 1967.

182. Stein, R.S., Rhodes, M.B., and Porter, R.S., *J. Colloid Interfac. Sci.*, 27, 336 1968.

183. Pincus, P., *Solid State Commun.*, 7, 415 1969.

184. Doane, J.W. and Johnson, D.L., *Chem. Phys. Lett.*, 6, 291 1970.

185. Blinc, R., Hogenboom, D.L., O'Reilly, D.E., and Peterson, E.M., *Phys. Rev. Lett.*, 23, 969 1969.

186. Weger, M. and Cabane, P., *Colloque Sur Les Cristaux Liquides,* Montpellier, France, June, 1969.

187. (a) Spence, R.D., Moses, H.A., and Jain, P.L., *J. Chem. Phys.*, 21, 380 1953;
     (b) Spence, R.D., Gutowsky, H.S., and Holm, C.H., *J. Chem. Phys.*, 21, 1891 1953.

188. Jain, P.L., Lee, J.C., and Spence, R.D., *J. Chem. Phys.*, 23, 878 1955.

189. Jain, P.L., Moses, H.A., Lee, J.O., and Spence, R.D., *Phys. Rev.,* 92, 844 1953.

190. Lippmann, H., *Ann. Phys.,* 2, 287 1958.

191. Lippmann, H. and Weber, K.H., *Ann. Phys.,* 20, 265 1957.

192. Weber, K.H., *Ann. Phys.,* 3, 1 1959.

193. Weber, K.H., *Discussions Faraday Soc.,* 25, 74 1958.

194. Saupe, A. and Englert, B., *Phys. Rev. Lett.,* 11, 462 1963.

195. Luckhurst, G.R., *Quart. Rev.,* 22, 179 1968.

196. Meiboom, S. and Snyder, L.C., *Science,* 162, 1337 1968.

197. Snyder, L.C. and Meiboom, S., *Mol. Cryst. Liquid Cryst.,* 7, 181 1969.

198. Diehl, P. and Khetrapal, C.L., in *NMR-Basic Principles and Progress,* Vol. 1, Springer-Verlag, Berlin, Germany, 1969, 1.

199. Saupe, A., *Z. Naturforsch.,* 19a, 161 1964.

200. Snyder, L.C., *J. Chem. Phys.,* 43, 4041 1965.

201. Carrington, A. and Luckhurst, G.R., *Mol. Phys.,* 8, 401 1964.

202. Rowell, J.C., Phillips, W.D., Melby, L.R., and Panar, M., *J. Chem. Phys.,* 43, 3442 1965.

203. Bravo, N., Doane, J.W., Arora, S.L., and Fergason, J.L., *J. Chem. Phys.,* 50, 1398 1969.

204. Saupe, A., *Mol. Cryst.,* 1, 527 1966.

205. Nehring, J. and Saupe, A., *Mol. Cryst. Liquid Cryst.,* 8, 403 1969.

206. Diehl, P. and Khetrapal, C.L., *Mol. Phys.,* 14, 283 1967.

207. Diehl, P. and Khetrapal, C.L., *J. Mag. Res.,* 1, 524 1969.

208. Yannoni, C., Second Symposium on Ordered Fluids and Liquid Crystals, American Chemical Society Meeting, New York, September 1969.

209. Carr, E.F., Hoar, E.A., and MacDonald, W.T., *J. Chem. Phys.,* 48, 2822 1968.

210. Diehl, P., Khetrapal, C.L., Kellerhals, H.P., Lienhard, V., and Niederberger, W., *J. Mag. Res.,* 1, 527 1969.

211. Sobajima, S., *J. Phys. Soc. Jap.,* 23, 1070 1967.

212. Sackmann, E., Meiboom, S., and Snyder, L.C., *J. Amer. Chem. Soc.,* 90, 2183 1968.

213. Carr, E.F., Parker, J.H., and McLemore, D.P., Second Symposium on Ordered Fluids and Liquid Crystals, American Chemical Society Meeting, New York, September 1969.

214. Flautt, T.J. and Lawson, K.D., in *Ordered Fluids and Liquid Crystals,* Advances in Chemistry Series, No. 63, Porter, R.S. and Johnson, J.F., Eds., American Chemical Society, Washington, D.C., 1967, Chapter 3.

215. Lawson, K.D. and Flautt, T.J., *J. Amer. Chem. Soc.,* 89, 5489 1967.

216. Black, P.J., Lawson, K.D., and Flautt, T.J., *Mol. Cryst. Liquid Cryst.,* 7, 201 1969.

217. Lawson, K.D. and Flautt, T.J., *J. Phys. Chem.,* 72, 2066 1968.

218. Fryberg, G.C. and Gelerinter, E., *J. Chem. Phys.,* in press.

219. Schwerdtfeger, C.F. and Diehl, P., *Mol. Phys.,* 17, 417 1969.

220. Diehl, P. and Schwerdtfeger, C.F., *Mol. Phys.,* 17, 423 1969.

221. Chen, D.H. and Luckhurst, G.R., *Trans. Faraday Soc.,* 65, 565 1969.

222. Chen, D.H. and Luckhurst, G.R., *Mol. Phys.,* 16, 91 1969.

223. Chen, D.H., James, P.G., and Luckhurst, G.R., *Mol. Cryst. Liquid Cryst.,* 8, 71 1969.

224. Glarum, S.H. and Marshall, J.H., *J. Chem. Phys.* 46, 55 1967.

225. Ferruti, P., Gill, D., Harpold, M.A., and Klein, M.P., *J. Chem. Phys.,* 50, 4545 1969.

226. Glarum, S.H. and Marshall, J.H., *J. Chem. Phys.,* 44, 2884 1966.

227. Falle, H.R. and Luckhurst, G.R., *Mol. Phys.,* 11, 299 1966.

228. Luckhurst, G.R., *Mol. Phys.,* 11, 205 1966.

229. Falle, H.R., Luckhurst, G.R., Lemaire, H., Marechal, Y., Rassat, A., and Rey, P., *Mol. Phys.,* 11, 49 1966.

230. Falle, H.R. and Luckhurst, G.R., *Mol. Phys.,* 12, 493 1967.

231. Luckhurst, G.R., *Mol. Cryst.,* 2, 363 1967.

232. Uhrich, D.L., Wilson, J.M., and Resch, W.A., *Phys. Rev. Lett.,* 24, 355 1970.

233. Tsvetkov, V.N. and Milhailov, G.M., *Soviet Phys. JETP,* 7, 1399 1937.

234. Mikhailov, G.M. and Tsvetkov, V.N., *Soviet Phys. JETP,* 7, 597 1939.

235. Miesovicz, M., *Nature,* 158, 27 1946.

236. Berherer, G. and Kast, W., *Ann. Phys.,* 41, 355 1942.

237. Fisher, J. and Fredrickson, A.G., *Mol. Cryst. Liquid Cryst.,* 8, 267 1969.

238. Sakamoto, K., Porter, R.S., and Johnson, J.F., *Mol. Cryst. Liquid Cryst.,* 8, 443 1969.

239. Porter, R.S., Barrall, E.M., and Johnson, J.F., *J. Chem. Phys.* 45, 1452 1966.

240. Porter, R.S. and Johnson, J.F., *J. Phys. Chem.,* 66, 1828 1962.

241. Helfrich, W., *Phys. Rev. Lett.,* 23, 372 1969.

242. Hoyer, W.A. and Nolle, A.W., *J. Chem. Phys.,* 24, 803 1956.

243. Zvereva, G.E., *Soviet Phys. Acoust.,* 11, 212, 1965.

244. Kapustin, A.P. and Bykova, N.T., *Soviet Phys. Cryst.,* 11, 297 1966.

245. Kapustin, A.P. and Zvereva, G.E., *Soviet Phys. Cryst.,* 10, 603 1966.

246. Dyro, J.F. and Edmonds, P.D., *Mol. Cryst. Liquid Cryst.,* 8, 141 1969.

247. Edmonds, P.D. and Orr, D.A., *Mol. Cryst.,* 2, 135 1966.

248. Zvereva, G.E. and Kapustin, A.P., *Akustichesku Z.,* 10, 122 1964.

249. Nordland, W.A., *J. Appl. Phys.,* 39, 5033 1968.

250. Durand, G. and Rao, D.G.L.N., *Phys. Lett.,* 7, 455 1968.

251. Cole, G.D., Merrit, W.G., and Walker, W.W., *J. Chem. Phys.,* 49, 1980 1968.

252. Cole, G.D., Univ. Microfilms 64-9117, *Dissertation Abstracts,* 25, 1967 1964.

253. Wesloyski, J., Szuszkiewicz, M., and Szuszkiewicz, S., *Acta Phys. Polon,* 29, 97 1966.

254. Chandrasekhar, S., Krishnamurti, D., and Madhusudana, *Mol. Cryst. Liquid Cryst.,* 8, 45 1969.

255. Lawrence, A.S.C., *Mol. Cryst. Liquid Cryst.*, 7, 1 1969.

256. Winsor, P.A., *Chem. Rev.*, 68, 1 1968.

257. Stewart, G.T., *Mol. Cryst.* 1, 563 1966.

258. Stewart, G.T., *Mol. Cryst. Liquid Cryst.*, 7, 75 1969.

259. Luzzati, V. and Reiss-Husson, F., *Nature,* 210, 1351 1966.

260. Luzzati, V., Mustacchi, H., Skoulios, A., and Reiss-Husson, F., *Acta Crystallogr.*, 13, 660 1960.

261. Luzzati, V. and Reiss-Husson, F., *Adv. Biol. Med. Phys.*, 11, 87 1967.

262. Ekwall, P., Mandell, L., and Fontell, K., *Mol. Cryst. Liquid Cryst.*, 8, 157 1969.

263. For example, McBain, J.W. and Lee, W.W., *Oil and Soap.*, 17 February 1943.

APPENDIX A

Since the compilation of this review on the structure and properties of liquid crystals, an international conference was held in Berlin, Germany. There were many interesting and stimulating papers presented at this conference. Most of these papers will be published as the proceedings of that conference and will appear first in the journal *Molecular Crystals and Liquid Crystals.* Following the journal publication, which will cover a number of issues, the proceedings of the conference will be brought together under a hardback binding. Because of the time factor involved in presenting all of the papers at the Berlin Conference it seemed appropriate that the titles of these papers be listed in the Appendix of this review. Along with the titles of the papers, we include the names and addresses of the authors.

Optical Properties

1. A Liquid Crystal Device for Rotating the Plane of Polarized Light

   John F. Dreyer
   Polacoat, Inc.
   9750 Conklin Road
   Cincinnati, Ohio 45242

2. Light Scattering Spectrum of a Nematic Liquid

   Ivan Haller
   IBM Corporation
   T.J. Watson Research Center
   Yorktown Heights, New York 10598

   J. D. Litster
   Department of Physics
   Massachusetts Institute of Technology
   Cambridge, Massachusetts

3. Scattering of Light by Liquid Crystalline p-Azoxyanisole

   D. Krishnamurti and
   H. S. Subramhanyam
   Department of Physics
   University of Mysore
   Mysore, India

4. Orientational Order in Anisaldazine in the Nematic Phase

   N. V. Madhusudana
   R. Shashidhar and
   S. Chandrasekhar
   Department of Physics
   University of Mysore
   Mysore, India

5. Spatial Distribution of Light Scattered by PAA in Applied Electric Field

   M. Bertolotti, B. Daino, F. Scudieri
   and D. Sette
   Fondazione U. Bordoni
   Instituto Superiore P. T.
   Roma, Italy
   and
   Istitute di Fisica della Facolta de Ingegnéria
   Universita di Roma

6. Liquid Crystal Matrix Displays Using Additional Solid Layers for Suppression of Parasitic Currents

   J. G. Grabmaier, W. F. Greubel, and
   H. H. Kruger
   Siemens Aktiengesellschaft
   Munich, West Germany

7. Radiography Utilizing Cholesteric Liquid Crystals

   J. L. Fergason
   Liquid Crystal Institute
   Kent State University
   Kent, Ohio 44242

   R. D. Ennulat
   U.S. Army Electronics Command
   Night Vision Laboratory
   Ft. Belvoir, Virginia 22060

8. The Selective Light Reflection by Plane Textures

   R. D. Ennulat
   U. S. Army Electronics Command
   Night Vision Laboratory
   Ft. Belvoir, Virginia 22060

9. Selective Reflection of Cholesteric Liquid Crystals

   R. Dreher and G. Meier
   Institut für angewandte
   Festkörperphysik der
   Fraunhofer-Gesellschaft
   78 Freiburg, W. Germany

   A. O. Saupe
   Liquid Crystal Institute
   Kent State University
   Kent, Ohio 44242

10. Temperature Resolution of the Human Eye by Means of Cholesteric Liquid Crystal Films

    G. Ittner and B. Böttcher
    Bundesanstalt für Materialprufung (BAM)
    1 Berlin 45, Unter den Eichen 87
    Germany

11. Optical Properties of Liquid Crystal Mixtures

    H. Stegemeyer and K. J. Mainusch
    Iwan N. Stranski - Institut für Physikalische Chemie
    der Technischen Universität
    West Berlin, Germany

12. Birefringence of Cholesteric Liquid Crystal Films
    B. Böttcher and G. Graber
    Bundesanstalt für Materialprüfung (BAM)
    1 Berlin 45, Unter den Eichen 87
    Germany

13. Optical Rotatory Dispersion of Cholesteric Liquid Crystals

    S. Chandrasekhar and J. Shashidara Prasad
    Department of Physics
    University of Mysore
    Mysore 6, India

14. Reflection and Transmission by Perfectly Ordered Cholesteric Liquid Crystal Films: Theory and Verification

    Dwight W. Berreman and Terry J. Scheffer
    Bell Telephone Laboratories, Incorporated
    Murray Hill, New Jersey

15. Birefringence of the Smectic Modification of Thallium Soaps

    H. Sackmann and G. Pelzl
    University of Halle
    Sektion Chemie
    DDR 402 Halle (S)
    Muhlpforte 1
    East Germany

16. Optical and Structural Effects in Type C Smectic Phases

    Ted R. Taylor, Sardari L. Arora
    and James L. Fergason
    Liquid Crystal Institute
    Kent State University
    Kent, Ohio 44242

Theory (Statistical or Continuum)

1. Molecular Theory of Nematic Liquid Crystals

S. Chandrasekhar and N. V. Madhusudana
Department of Physics
University of Mysore
Mysore, India

2. On the Validity of the Maier-Saupe Theory of the Nematic Transition

   Theodore D. Schultz
   IBM Research Center
   Yorktown Heights, New York 10598

3. Successive Transitions in a Nematic Liquid

   M. J. Freiser
   IBM Watson Research Center
   Yorktown Heights, New York 10598

4. The Statistical Mechanics for Long Semi-Flexible Molecules: A Theory of the Nematic Mesophase

   Andrew G. De Rocco and Alexander Wulf
   Department of Physics and Astronomy
   University of Maryland
   College Park, Maryland 20742

5. NENDORECS* in Liquid Crystals [*Nuclei with an Equal Number of Degrees of Order Resonating in an Electron Cloud]

   W. P. Holland
   "Thurnham"
   Egremont Road
   Hensingham
   Whitehaven
   Cumberland, England

6. Fluid Phases of Rigid Molecules of High Asymmetry

   L. K. Runnels and Carolyn Colvin
   Department of Chemistry
   Louisiana State University
   Baton Rouge, Louisiana

7. Scaled Particle Theory of Two Dimensional Anisotropic Fluids

   K. Timling
   Philips Research Laboratories
   N. V. Philips' Gloeilampenfabrieken
   Eindhoven, Netherlands

8. Experimental and Theoretical Studies of Alignment Singularities in Liquid Crystals

   Jurgen Nehring
   Liquid Crystal Institute
   Kent State University
   Kent, Ohio 44242

9. A Molecular Theory of The Cholesteric Phase

   W. J. A. Goossens
   Philips Research Laboratories
   N. V. Philips' Gloeilampenfabrieken
   Eindhoven, Netherlands

10. An X-Ray Optical Examination of Binary Cholesteric Liquid Crystal System

    M. A. Khan and D. J. Morantz
    Woolwich Polytechnic
    Wellington Street
    London, England

11. A New Elastic-Hydrodynamid Theory of Liquid Crystals

    P. C. Martin and Jack Swift
    Department of Physics and
    P. S. Pershan
    Division of Engineering and Applied Physics
    Harvard University
    Cambridge, Massachusetts

NMR and EPR

1. Spin Relaxation in Nematic Liquid Crystals and in Solutes in Liquid Crystals

   J. W. Doane, J. A. Murphy and
   J. J. Visintainer
   Department of Physics and
   Liquid Crystal Institute
   Kent State University
   Kent, Ohio 44242

2. Thermal Fluctuations and Proton Spin-Lattice Relaxation in Nematic Liquid Crystals

Assis Farinha-Martins
Laboratoire de Resonance Magnetique
Centre d' Etudes Nucleaires de Grenoble
Cedex 85—38 Grenoble-Gare
France

3. Nuclear Quadrupolar Relaxation of N^{14} in a Nematic Liquid

B. Cabane and Gilbert Clark
Universite de Paris
Faculte des Sciences
Service de Physique des Solides
Batiment 510
91 - Orsay
France

4. Geometrical Structure Data of $1-{}^{13}$C-Ethylene

H. Spiesecke
Euratom
CCR
Ispra, Italy

5. NMR Spectra of 1,2-Difluorobenzene in Nematic Solvents. Anisotropy of Indirect Fluorine Couplings and Molecular Geometry

J. Gerritsen and C. MacLean
Scheikundig Laboratorium
Der Vrije Universiteit
Amsterdam - Z

6. Measurement of Deuterium Nuclear Quadrupole Coupling Constants in Liquid Crystal Solutions

B. M. Fung and I. Y. Wei
Tufts University
Medford, Massachusetts

7. NMR Study of the Order Parameter in a Nematic Phase at Room Temperature

J. P. Le Pesant and P. Papon
Faculte des Sciences de Paris
Laboratoire de Resonance Magnetique
9, Quai St-Bernard
Paris 5e, France

8. Self-Diffusion, Spin Relaxation and Long Range Order in Liquid Crystals

R. Blinc, J. Pirs, M. Vilfan, I. Zupančič and V. Dimic
The University of Ljubljana
Institute "J. Štefan"
Ljubljana, Yugoslavia

9. Liquid Crystal Ordering in the Magnetic and Electric Fields as Studied in 4,4′-Bis Heptyloxyazoxybenzene by Electron Paramagnetic Resonance

M. Šentjurc and M. Schara
Institute "J. Štefan"
Ljubljana, Yugoslavia

10. EPR Study of the Temperature Dependence of Molecular Rotation in a Nematic Liquid Crystal

C. F. Schwerdtfeger and M. Marušič
Physics Department
University of British Columbia
Vancouver, Canada

11. Counter-Ion Binding in Some Lyotropic Liquid Crystalline Phases Studied by NMR

Goran Lindblom and Bjorn Lindman
Division of Physical Chemistry 2
The Lund Institute of Technology
P.O.B. 740, S-220 07 Lund 7
Sweden

12. Deuterium and Proton Magnetic Resonance Investigation of Lamellar Lyotropic Liquid Crystals

Ake Johansson and Torbjorn Drakenberg
Department of Physical Chemistry
The Lund Institute of Technology
Lund, Sweden

13. NMR Study of Molecular Motions in the Mesophases of Potassium Laurate - D_2O System

J. Charvolin and P. Rigny
Laboratoire de Physique des Solides
Faculte des Sciences
91 - Orsay, France

Polymorphism and Structure

1. On the Existence of Nematic Secondary Structures

   Ludwig Pohl and Ralf Steinstraesser
   Research Laboratories
   E. Merck
   Darmstadt, West Germany

2. X-Ray Studies of Some Smectic, Nematic, and Isotropic Phases

   Adriaan de Vries
   Liquid Crystal Institute
   Kent State University
   Kent, Ohio 44242

3. An X-Ray Study of the Mesophases of Cholesteryl Stearate

   John E. Lydon
   The University of Leeds
   The Astbury Department of Biophysics
   Leeds 2, England

4. Holomicroscopy of Liquid Crystals

   M. B. Rhodes
   Department of Chemistry
   University of Massachusetts
   Amherst, Massachusetts 01003

5. X-Ray Diffraction and Polymorphism of Smectic Liquid Crystals

   S. Diele, H. Sackmann and P. Brand
   Universität Halle
   Sektion Chemie
   DDR 402 Halle (S)
   Mühlpforte 1
   East Germany

6. Investigation of a Smectic Tetramorphous Substance

   D. Demus, S. Diele, M. Klapperstück, V. Link and H. Zaschke
   Universität Halle
   Sektion Chemie
   DDR 402 Halle (S)
   Mühlpforte 1
   East Germany

7. Liquid Crystalline Structures for Binary Systems $\alpha - \omega$ Soaps/Water. Electric Interactions in These Systems

   Bernard GALLOT
   C.B.M.
   45-Orléans 02
   France

8. X-Ray Diffraction Studies on Oriented Lyotropic Mesomorphic Phases

   R. R. Balmbra and J. S. Clunie
   Basic Research Department
   Procter & Gamble Limited
   Newcastle Technical Centre
   Newcastle upon Tyne, NE12 9 TS.
   England

9. Nematic Order and Structure in Amorphous Polymers

   G. S. Y. Yeh
   Department of Chemical and
   Metallurgical Engineering
   University of Michigan
   Ann Arbor, Michigan

10. Study of Liquid-Crystalline Structures of AB and ABA Polybutadiene - Polystyrene Block Copolymers, by Small Angle X-Ray Scattering and Electron Microscopy

    Bernard Gallot and André Douy
    C.B.M.
    45 - Orléans 02
    France

11. Phase Diagram of Systems: Block-Copolymer/Preferential Solvent of One Block

    Bernard Gallot, Monique Gervais and André Douy
    C.B.M.
    45-Orléans 02
    France

12. Polymorphism in Cholesteryl Esters: Cholesteryl Palmitate

Marcel J. Vogel, Charles P. Mignosa
IBM Advanced Systems Development Division Lab.
Edward M. Barrall II
IBM Research Laboratory
San Jose, California 95114

Chemistry of Liquid Crystals

1. The Liquid Crystalline Behaviour of Alkyl, Aryl and Arylalkyl 4-p-Substituted Benzylideneamino-Cinnamates and -α-Methylcinnamates

   G. W. Gray and K. J. Harrison
   Department of Chemistry
   The University of Hull
   Yorkshire, England

2. Nematic Liquids as Reaction Media for the Claisen Rearrangement

   W. E. Bacon and Glenn H. Brown
   Liquid Crystal Institute
   Kent State University
   Kent, Ohio 44242

3. The Conformation of Aromatic Schiff Bases in Connection with Liquid Crystalline Properties

   J. van der Veen
   Philips Research Laboratories
   N. V. Philips Gloeilampenfabrieken
   Eindhoven, Netherlands

4. Molecular Structure and Mesomorphic Properties of Phenylenealkoxybenzoates

   Sardari L. Arora, James L. Fergason
   and Ted R. Taylor
   Liquid Crystal Institute
   Kent State University
   Kent, Ohio 44242

5. Liquid Crystals V. Room Temperature Nematic Materials Derived From p-Alkylcarboanto-p'-Alkoxyphenyl Benzoates

   Joseph A. Castellano, Michael T. McCaffrey and Joel E. Goldmacher
   RCA Laboratories
   David Sarnoff Research Center
   Princeton, New Jersey 08540

6. The Odd-Even Effect in Steryl ω-Phenylalkanoates

   W. Elser, J. L. W. Pohlmann and P. R. Boyd
   U.S. Army Electronics Command
   Night Vision Laboratory
   Ft. Belvoir, Virginia 22060

7. Structure Dependence of Cholesteric Mesophases II, III, & IV

   J. L. W. Pohlmann and W. Elser
   U.S. Army Electronics Command
   Night Vision Laboratory
   Ft. Belvoir, Virginia 22060

8. Mesomorphism of Homologous Series II. Odd-Even Behavior

   R. D. Ennulat and A. J. Brown
   U.S. Army Electronics Command
   Night Vision Laboratory
   Ft. Belvoir, Virginia 22060

9. Mesomorphic Substances Containing Some Group IV Elements

   William R. Young, Ivan Haller
   and Dennis C. Green
   IBM Watson Research Center
   Yorktown Heights, New York 10598

10. Mesomorphic Behaviour of Schiff's Base Compounds-I: N-N'-DI (4-n-Alkoxy-1-Naphthylidine) -Benzidines

    J. S. Dave, A. P. Prajapati
    and R. A. Vora
    Chemistry Department
    Faculty of Science
    M. S. University of Baroda
    Baroda, India

11. The Liquid Crystalline State Copolymerisation

    A. A. Baturin, Y. B. Amerik
    and D. A. Krentsel
    Topchiev Institute of Petrochemical Synthesis of Academy of Science
    Moscow, U.S.S.R.

12. Polymerization of Nematic Liquid Crystal Monomer

    C. M. Paleos and M. M. Labes
    Department of Chemistry
    Drexel University
    Philadelphia, Pennsylvania 19104

13. Polymerization of Methacryloyloxy Benzoic Acid Within Liquid Crystalline Media

    A. Blumstein, N. Kitagawa
    and R. Blumstein
    Department of Chemistry
    Lowell Technological Institute
    Lowell, Massachusetts

14. Comparison of the Radiolytically Induced Polymer Formation in Alkenyl Carboranes in the Mesomorphic State and the Crystalline State

    T. J. Klingen and J. R. Wright
    Department of Chemistry
    The University of Mississippi
    University of Mississippi

15. Mesomorphic Behaviour of Cholesteryl Esters II: Trans-p-n-Alkoxycinnamates of Cholesterol

    J. S. Dave and R. A. Vora
    Chemistry Department
    M. S. University of Baroda
    Baroda, India

Living Systems

1. Principles Underlying 'Fluctuating Crystallinity' in Cell Membranes and Other Biological Structures and Its Implications

    R. K. Mishra, A. de Vries,
    G. H. Brown and E. J. Petscher
    Liquid Crystal Institute
    Kent State University
    Kent, Ohio 44242

    S. Barenberg and P. H. Geil
    Institute of Polymer Science
    Case Western Reserve University
    Cleveland, Ohio 44106

    William H. Falor
    Lymph Research Laboratory
    Akron City Hospital
    Akron, Ohio

2. Binary Mixtures of MBBA and Rod-Like Molecules

    P. E. Cladis, J. Rault and J. P. Burger
    Laboratoire de Physique des Solides
    Faculte des Sciences
    91 Orsay, France

3. Antibiotic Polymers

    B. T. Butcher, M. K. Stanfield, G. T. Stewart
    S. S. Wagle and R. Zemelman
    Tulane University Medical Center
    New Orleans, Louisiana 70112

External Field Effects

1. A.C. and D.C. Regimes in the Electrohydrodynamical Unstabilities of a Nematic Liquid Crystal

    Orsay Liquid Crystal Group
    Faculte des Sciences
    91 Orsay, France

2. Ordering in the Smectic Phase Owing to Electric Fields

    E. F. Carr
    Physics Department
    University of Maine
    Orono, Maine 04473

3. Continued Kinetic Study of the Cholesteric $\rightleftarrows$ Nematic Transition in a Liquid Crystal Film

    J. J. Wysocki
    Corporate Exploratory Development Laboratory
    Xerox Corporation
    Rochester, New York

4. Structural Effects in Cholesteric Liquid Crystals

    C. J. Gerritsma and P. V. Zanten
    Philips Research Laboratories
    N. V. Philips' Gloeilampenfabrieken
    Eindhoven, Netherlands

5. Dipole Relaxation in a Liquid Crystal

   H. Baessler, R. B. Beard and M. M. Labes
   Department of Chemistry and Electrical Engineering
   Drexel University
   Philadelphia, Pennsylvania 19104

6. Distortion of Orientation Patterns in Liquid Crystals by a Magnetic Field

   F. M. Leslie
   Mathematics Department
   University of Strathclyde
   Glasgow, Scotland

7. Effects of Electric Fields on Nematic Liquid Crystals

   J. Nemec and B. D. Cook
   Department of Electrical Engineering
   University of Houston
   Houston, Texas 77004

8. Correlation Between Electrical Properties of Optical Behaviour of Nematic Liquid Crystals

   H. Gruler and G. Meier
   Institut fur Angewandte
   Festkorperphysik der
   Fraunhofer-Gesellschaft
   78 Freiburg, West Germany

9. Etude Du Courant Dans Les Cristaux Liquides Nematiques

   Georges Assouline, Michele Hareng
   and Eugene Leiba
   Thomson-CSF
   Domaine de Corbeville
   91 Orsay, France

10. Achievement of a Pronounced Threshold in the Dynamic Scattering of Nematic Liquid Crystals

    Ulrich Wolff
    Siemens Aktiengesellschaft
    Zentrale Forschung und Entwicklung
    Forschungslaboratorien
    8 Munchen 25, Postfach 700
    West Germany

11. Electro-Optic Effect in a Room Temperature Nematic Liquid Crystal

    W. Haas, J. Adams and J. B. Flannery
    Xerox Research Laboratories
    Webster, New York 14580

12. On Pyroelectrical Effects in Liquid Crystals

    A. Szymanski
    Institute of Physics
    Technical University of Lodz
    Poland

13. Electric Field Induced Vorticity in p-Azoxyanisole

    P. Andrew Penz
    Scientific Research Staff
    Ford Motor Company
    Dearborn, Michigan 48121

14. Point Disclinations at a Nematic-Isotropic Liquid Interface

    Robert B. Meyer
    Harvard University
    Division of Engineering and Applied Physics
    Pierce Hall
    Cambridge, Massachusetts 02138

15. Electrohydrodynamic Flow in Nematic Liquid Crystals

    H. Koelmans and A. M. van Boxtel
    Philips Research Laboratories
    N. V. Philips' Gloeilampenfabrieken
    Eindhoven, Netherlands

16. Ionic Equilibrium and Ionic Conductance in the System Tetra-*Iso*-Pentyl Ammonium Nitrate - p-Azoxyanisole

    Alan Sussman
    RCA Laboratories
    Princeton, New Jersey 08540

17. Cholesteric - Nematic Phase Transitions

    G. R. Luckhurst and H. J. Smith
    Department of Chemistry
    University of Southampton

Southampton, S09 5NH
England

18. Electrohydrodynamic Phenomena in Connection with Unipolar Injection of Charge Carriers in M.B.B.A. Deionized by Electrodialysis

   Tobazeon, Filippini (C.N.R.S.)
   Borel, Robert, Poggi (L.E.T.I.)
   Groupe d'Etudes des Cristaux Liquides
   LETI-CNRS
   France

19. Effect of Pressure on the Mesomorphic Transition in Para-Azoxyanisole (P.A.A.)

   B. Deloche, B. Cabane, D. Jerome
   Laboratoire de Physique des Solides
   Faculte des Sciences
   91 Orsay, France

Physical Properties of Liquid Crystals

1. Characterization of Molecular Role in Pitch Determination in Liquid Crystal Mixtures

   J. Adams and W. Haas
   Xerox Research Laboratories
   Webster, New York 14580

2. Effects of Solutes on the Helical Twist of Cholesteric Liquid Crystals

   H. Baessler, I. Teucher and M. M. Labes
   Department of Chemistry
   Drexel University
   Philadelphia, Pennsylvania 19104

3. Cholesteric Texture Near T_c and in Presence of a Magnetic Field

   J. Rault and P. E. Cladis
   Laboratoire de Physique des Solides
   Faculte des Sciences
   91 Orsay, France

4. NMR Studies of Selected Cholesteric Compounds

   L. M. Cameron and R. E. Callender
   U.S. Army Electronics Command
   Night Vision Laboratory
   Ft. Belvoir, Virginia 22060

5. A Lattice Model for Crystal-Nematic and Nematic-Isotropic Liquid Transtitions

   John T. S. Andrews
   Liquid Crystal Institute
   Kent State University
   Kent, Ohio 44242

6. Ultrasonic Absorption Measurements in Liquid Crystals

   S. Candau and P. Martinoty
   Universite de Strasbourg
   Faculte des Sciences
   Institute de Physique
   3, rue de l'Universite
   Strasbourg, France

7. Torques in Sheared Nematic Liquid Crystals: A Simple Model in Terms of the Theory of Dense Fluids

   W. Helfrich
   RCA Laboratories
   Princeton, New Jersey 08540

8. Theory of Diffusion in Liquid Crystal Mesophases

   Wilbur M. Franklin
   Kent State University
   Kent, Ohio 44242

9. Considerations on the Surface Tension and Viscosity Anomalies of Liquid Crystals

   Bun-ichi Tamamushi
   Nezu Chemical Institute
   Musashi University
   Tokyo, Japan

10. Viscosity Measurements of Nematic Liquid Crystal by a Shear Waves Reflectance Technique

    P. Martinoty and S. Candau
    Universite de Strasbourg

Faculte des Sciences
Institut de Physique
3, rue de l'Universite
Strasbourg, France

11. Anisotropy of Self-Diffusion in a Liquid Crystal Studied by Neutron Quasielastic Scattering

J. A. Janik, J. M. Janik, K. Otnes
and T. Riste
Institutt for Atomenergi
Kjeller, Norway

12. Heat Transport in Liquid Crystals

J. O. Kessler and M. T. Longley-Cook
Physics Department
University of Arizona
Tucson, Arizona

13. Theory of Translational and Orientational Melting with Application to Liquid Crystals

Kenji K. Kobayashi
Institute for Solid State Physics
University of Tokyo
Roppongi, Tokyo

14. Gas-Liquid Chromatography Determination of the Degree of Order in a Nematic Mesophase

Laurence C. Chow and Daniel E. Martire
Department of Chemistry
Georgetown University
Washington, D. C. 20007

15. New Applications of Nematic Phases in Gas-Liquid-Chromatography

H. Kelker, J. Jainz, J. Sabel
and H. Winterscheidt
Farbwerke Hoechst A.G.
Frankfurt/M.80

16. The Preparation and Thermodynamics of Some Homologous Nitrones, A New Group of Liquid Crystals

William R. Young, Ivan Haller
and Arieh Aviram
IBM Watson Research Center
Yorktown Heights, New York 10598

APPENDIX B

Diffusion

During the writing of this review, the topic of diffusion was omitted since the literature was nearly devoid of such studies. However, several papers have recently appeared which are interesting and in some cases provide surprising results.

1. Blinc, R. and Dimic, V., *Phys. Lett., 31A*, 531, 1970.

2. Blinc, R., Easwaran, K., Pirš, j., Volfan, M., Zupančič, I., *Phys. Rev. Lett.*, 25, 1327, 1970.

3. Murphy, J. A. and Doane, J. W., *Mol. Cryst. Liq. Cryst.*, to be published.

4. Yun, C. and Fredrickson, A. G., *Mol. Cryst. Liq. Cryst.*, to be published.